The Age of Synthesis

Worcester Polytechnic Institute
Studies in Science, Technology and Culture

Lance Schachterle and Francis C. Lutz
Co-Editors

Vol. 16

PETER LANG
New York • Washington, D.C./Baltimore • San Francisco
Bern • Frankfurt am Main • Berlin • Vienna • Paris

Carl W. Hall

The Age of Synthesis

A Treatise and Sourcebook

PETER LANG
New York • Washington, D.C./Baltimore • San Francisco
Bern • Frankfurt am Main • Berlin • Vienna • Paris

Library of Congress Cataloging-in-Publication Data

Hall, Carl W.
The age of synthesis: a treatise and sourcebook / by Carl W. Hall.
p. cm. — (Worcester Polytechnic Institute studies in science, technology and culture; vol. 16)
Includes bibliographical references and index.
1. Technology—Philosophy. 2. Technology—Social aspects. I. Title. II. Series.
T14.H285 601—dc20 94-13698
ISBN 0-8204-2455-2
ISSN 0897-926X

Die Deutsche Bibliothek-CIP-Einheitsaufnahme

Hall, Carl W.:
The age of synthesis: a treatise and sourcebook / Carl W. Hall. - New York; Washington, D.C./Baltimore; San Francisco; Bern; Frankfurt am Main; Berlin; Vienna; Paris: Lang.
(Worcester Polytechnic Institute studies in science, technology and culture; 16)
ISBN 0-8204-2455-2
NE: Worcester Polytechnic Institute: Worcester Polytechnic Institute . . .

SYNTHESIS

is a way of thinking and doing,
of providing a vision, in which an idea or a thing,
imagined or real, is seen as a coherent whole;
often consisting of parts, from which a thought can be developed,
action can be rejected or taken, and the thing made, assembled, or constructed; either as a new creation or activity or as a duplicate
or substitute of known substances.

unity

holistic

systems

integration

systemization

interconnectivity

ACKNOWLEDGEMENTS

Win Aung, Greg Barthold, Fred Betz, Kris Boone, Leo Bustad, J. A. Davis, A. J. Eggenberger, Thomas O. Eisemon, Claudia Genuardi, Mary J. Golladay, Robert Gustafson, George Hazelrigg, Chris Hill, Lauritz Holm-Nielsen, Thomas P. Hughes, Sue Kemnitzer, Louis Martin-Vega, Henry A. McGee, Ohio State University Graduate School, Morio Okazaki, William A. Owczarski, John Person, George Peterson, Lynn Preston, Decatur Rogers, John Scalzi, Ruth Slonim, N. P. Suh, Walter Vergara, John A. White, and Kenneth G. Wilson.

Services of Libraries at The Ohio State University, the George Washington University (Gelman), the George Mason University (Law), the National Science Foundation were particularly helpful. The Arlington County Library was always close by and accessible for reference material.

Appreciation is expressed to Karla Kalasz, Pullman, WA 99163, who did the figures and sketches using computer graphics, to Amy Vincent for the portrait sketches, and to Christopher Valenti for electronic composition of all the elements.

Parts (Contents)

Matrixes (Tables) and Images (Figures)

LIST OF MATRIXES

LIST OF IMAGES

Grande Disco, A Sculpture in Charlotte, NC.

Information regarding ***Grande Disco*** *by the artist:*

Our life today is one of crisis…of movement…of tension. We do not know what our world will become. I try to say something about this uncertainty in my work. I try to communicate a sense of vitality and connection with the movement of life today…and to be a part of its movement.

The social challenge of art today, in my opinion, is to start a dialogue with the people.

I hope that is what happens with the **Grande Disco**.

Arnaldo Pomodoro
October 2, 1974
Charlotte, NC

Part 0.
Beginning

PREFACE

The **Age of Synthesis** is directed to those people interested in interdisciplinary, multidisciplinary, and crossdisciplinary studies and work in the professions and applied fields, policy development, and education. The objective is to show the increasing importance of synthesis in philosophical thinking and technological approaches to meet the needs of the 21st Century. The synthesis approaches are built on the strong analysis that developed over the past centuries, particularly the last century in Western culture. Educational programs, particularly at the university level, need to be revamped to incorporate a better understanding of the role of synthesis to serve the future needs of mankind.

The synthesis approaches provide a basis of improving the understanding between the arts and sciences, demonstrate the similarity of the professions and applied fields, and provide direction to identify gaps in knowledge.

This monograph, **The Age of Synthesis**, is based primarily on Western philosophy and technology. Interesting contrasts could undoubtedly be made with the experiences and results of Eastern philosophy. Eastern philosophy incorporates a greater concern for the whole, or holistic, not only of the person but also of the world surrounding the person. The philosophy of Native American culture is based on a synthesis of humans and Nature. The Eastern philosophies provide examples of an emphasis on synthesis, but do not have the extensive analysis that is a part of the Western philosophies.

The literature on synthesis is not nearly as extensive as that for analysis. Indexes, book contents, encyclopedias, and dictionaries often identify or list synthesis as a part of analysis, providing a limiting approach. Several books which are listed in references use the word synthesis in their titles. These books have been helpful in preparing this monograph. In the future more visibility will be given to synthesis, and perhaps a philosophy of synthesis will develop based on its own meaning and merits, not necessarily tied to analysis. I am not deterred by Medawar's (1967) statement, " ...of the use of those tiresome adjectives synthetic and analytic," and a person who uses synthesis in the sense of deductivity, not a common view.

The definition of words by philosophers is often different than the general usage. I have taken the liberty of taking words used by others, in Aristotle's writings regarding *a priori* and *a posteriori* , and in Kant's writings for analysis and synthesis, and adapting their concepts and words to today's situation. Analytic-synthetic was introduced by Kant and is used by many authors and often interpreted somewhat differently. Whitehead uses the word 'eternal' to mean existing forever, and although the dictionary meaning is the same, substitution of the word 'external' helps clarify its use in terms of synthesis by non-philosophers.

The words induction, deduction, intuition, and inference, and the combination of words as inductive inference and intuitive induction are subject to many different interpretations. Lacey (1986) gives several examples. Deduction was used more loosely by early philosophers. Intuition is now generally thought of as a direct relation between mind and some object without senses, whereas Kant used intuition in relations with sensible objects, more like deduction. In this monograph the recent generally accepted dictionary meanings are used.

The use of analysis-synthesis provides a common philosophical base applicable to arts, sciences, and applied fields, whereas the use of inductive-deductive is primarily the base for sciences. The likelihood of strengthening the interrelationships among all fields of knowledge is vastly improved by using the analysis-synthesis approach. Synthesis is now less strongly developed than analysis and needs to be inculcated in our future development. Synthesis can be applied to the thought process and the creation and growth of natural systems as well as the production of artificial systems. The development of synthesis as a way of thinking is paramount and teaching of synthesis should be incorporated in educational systems. A comprehensive definitional and historical appendix is included to assist those far removed from the subject. These materials should help the reader to develop a background on analysis and synthesis.

My approach has been to develop a view of synthesis pertinent to current events, then read and discuss what others think, followed by adjusting those views. This approach is in contrast to accepting the views of the well-known philosophers conceived in the past as a first step.

I do not claim to be a philosopher in the scholarly or academic sense.

"Philosophy means looking at the whole of a question, without reservations on simplifications; looking at ends and purposes, not merely at methods and means, and scrutinizing the latter in light of the former." Thomson (1954)

This statement fits my view of philosophy that would be rejected by some philosophers and scientific scholars, particularly the aspect of 'looking at ends and purposes.'

"I am sorry, then, that I pretended to be a philosopher." Hume

The Author.

THEME

» analysis needs to be strong and critical and is important for a strong synthsis

» analysis and synthesis are complementary, although often portrayed as opposites

» synthesis is a way of thinking as well as a way of doing, of understanding things, of making things, and can be used as a guide for analysis

» integration leads to a system approach, an exemplification of synthesis

» professions have a responsibility of emphasizing synthesis in their work/activities and must be educated to understand their role and how to carry out their work in an ethical manner

» validation of synthesis involves using critical analysis, meeting user needs, involving panels, teams and interdisciplinary groups to consider matters

» use of the approach of combined analysis-synthesis to replace induction-deduction to reduce the conflict of the two cultures and to build a unity

SYNOPSIS

Synthesis will supplement analysis as a major thrust of technological society. The present century is widely recognized as the Age of Analysis both philosophically and technologically. *A posteriori* evidence is now accumulating to demonstrate that the next century will be the Age of Synthesis. Synthesis will involve not only the substitution of materials but the building of new systems, meeting the needs and wants of society, and responding to imaginative thinking. Synthesis will help give direction to analysis by identifying gaps in information. So the trend will be to synthesis-analysis as contrasted to the present approach based on analysis-synthesis and the deductive-inductive approach used by scientists.

The philosophic foundations have been laid for synthesis, beginning with the early Greek philosophers, stimulated by the writings of I. Kant, and evidenced in the thinking of recent philosophers and demonstrated in the arts, sciences, and professions. Kant presented the philosophy of analytic and synthetic thinking of language along with the *a priori* and *a posteriori* concepts.

Philosophy has focused mainly on the mental the natural domains. Synthesis is a valuable concept for the unnatural or artificial (built, constructed, engineered, or manufactured devices and systems) domain. The arts utilize a predominance of the synthetic approach; the sciences a predominance of the analytic approach, with the notable exception of physics; and the professions a predominance of the synthetic approach.

The analytic approach, as represented in Western culture, has served the development of technological society well over the past 100 yr. and continues to do so. Without a strong analytic approach technological society would not have become as pervasive as has occurred. The synthetic approach incorporating the spirit and nature is represented by some Eastern cultures, but has lacked the analytic approach. In the future synthesis will be of major importance in Western cultures and Eastern cultures will incorporate more analysis as is evidenced by recent trends and will give a more holistic approach to both.

Scholars are involved in a search for truth (science) or beauty (art). The search for truth utilizes several approaches: deduction and induction; tuition and intuition; syllogism; and dialectical discussion; as well as analysis and synthesis. These approaches have been cloaked in such movements as dogmatism, existentialism, idealism, materialism and mechanism, phenomenology, positivism, pragmatism, rationalism, transcendentalism, and utilitarianism. These approaches and processes used to search for truth have served well in the past but need to be considered in a new matrix to more effectively meet the needs of the future. The emphasis has been primarily on analysis with secondary consideration for synthesis. There has been a rapid increase over the past quarter of a century in the use of synthesis, particularly in the physical world, and the stage is set for major efforts to emerge based on synthesis.

Leaders in practically all fields of arts and sciences and professions are suggesting a major increase in synthesis activity. The philosophy of synthesis provides a matrix or framework for meeting human needs through technological processes, while reaching for what might be as well as what is. The two cultures syndrome described so vividly by C. P. Snow would be minimized if the arts and sciences would use the same matrix (analysis-synthesis) for discussion rather than giving dominance to the science approach (deductive-inductive).

In thinking synthetically one needs to develop a more global view, consider a larger system, and often involve human, natural and physical systems simultaneously. Often synthetic thinking is done on an 'order of magnitude' scale, whereas analysis is usually based on precision. Inductive and deductive approaches do not provide that kind of matrix for thinking, but are an important part of the thinking process. Rules, axioms, or postulates are needed to guide and teach synthetic thinking. A start could be provided using some of the theorems developed for systems theory and rules for hierarchial systems. Systems are characterized by connectivity, interfacing, interactions, complexity, interdependence, and integration all of which relate to synthesis.

The validation of synthesis is more difficult than for analysis. There are possibilities for 'mischief' by those who would be too aggressive in promoting synthesis, primarily because of the difficulty in repeating the process in an attempt to get the same answer, which is not possible. Synthesis must stand the rigors of analysis, meet the needs of the user, and for those in the professions, should operate within an accepted code of ethics.

Synthesis is a pattern of nature. As more is learned about natural systems, that knowledge will be used not only to change those systems (genetic engineering, for example) but also to mimic for developing artificial systems (artifacts). People in the professions, people who design (in the broad sense), architects, engineers, veterinarians, and medical doctors need a stronger understanding of the role of synthesis in their work. New fields are developing within disciplines, such as synthetic chemistry, that require an entirely new approach. The Nobel Prize is now being used to recognize people who are synthesists (CATScan, DNA-double helix, economics, for example) in contrast to the tradition of recognizing analysts.

Most of the disciplines have been built on analysis, reductivism, and particularization rather than on synthesis, integration, and generalization. The growth of the interdisciplinary or multidisciplinary fields (sometimes called the hyphenated fields such as biochemistry, biophysics) may be the result of the established fields failing to change to incorporate the synthesis methods and to make a place for new fields.

In philosophy, including metaphilosophy and metaphysics, thinking has moved beyond pure thoughts to consideration for human concerns, natural systems, and artificial systems, more or less in this order. In the early years philosophers were the thought leaders, and they might also be involved in other pursuits such as in the arts, architecture, engineering, law, medicine, and the sciences.

THOUGHTS (ideas, imagination) become tangible as THESES (words, sketches, designs, treatises), which is a step to providing artifacts of that thought process as THINGS (art, manufactured goods, systems), represented by

THOUGHTS – to – THESES – to – THINGS

The flow from THOUGHTS to THINGS is synthesis, often with a tangible activity between. Nature provides the innate driving force for natural things and end use is the driving force for artificial things. Preceding the resulting artifact someone must conceive and have a vision of the end product. That conception occurs in the thought process of synthesis using deduction, induction, intuition, and creativity.

Synthesis is more useful than analysis for anticipating future developments, incorporating a larger sphere of participants and subjects in, and for setting the agenda for the next century. More effort should be devoted to preparing people to participate and to contribute in the world of synthesis. Even architecture and engineering that have design and therefore are involved in synthesis need to consciously strengthen the methods of synthesis in teaching and practice. Team and group activities will take on greater importance, more disciplines will be involved, and larger and more complex systems will be studied and built. Scholarly, technical and professional societies need to make a place for activities of synthesis on committees, in journals, and for presentations. There will be a shift from analysis-synthesis, which has been primarily deductive-inductive, to synthesis-analysis.

Part 1.
Introduction

The main objective of this treatise is to illustrate that we are in the beginning stages of a new age of synthesis, philosophically and technologically, as we enter the 21st Century. The major forcing function is the consumer demand for manufactured products, processed foods, and mass works of art. It is expected that these demands will be met without negatively impacting the environment, using a minimum of energy, and under safe conditions. This new age of synthesis is being built on and utilizes the emphasis on analysis of the 20th Century, just as analysis was built on the thinking of previous centuries. There is not a sharp distinction identifying the beginning of the age of synthesis that can better be determined on the basis of the record and events after the age of synthesis has developed. Synthesis as used in the philosophical sense has been with us for centuries, but has been mostly dormant in the recent century. Synthesis is an important activity in science, but has not been generally accepted, except after analysis. Synthesis is now becoming alive and will take its place beside analysis as an important approach, particularly for unnatural and constructed systems. Educational systems need to incorporate instruction and practice in synthesis to better prepare human resources for the future.

For the purpose of this presentation the methods of thinking, reasoning, and dialectical argument will focus on the two categories of analytic/analysis and synthetic/synthesis. Synthesis is used in the broadest sense, as a way of thinking as well as for the substitution or replacement or imitation of materials and systems, and for the development of new systems. The development of new materials and systems is emphasized, that is the artificial or artifacts or unnatural in contrast to the natural. Contrasts and similarities of natural and unnatural or artificial systems will be used as a basis of comparison and explanation.

Synthesis is becoming ever more useful and applicable as an approach to consider not only for larger artificial systems with more components involving more complexity but to consider human systems, either alone or as a part of unnatural or artificial systems. Examples of such systems in which human elements, in addition to technological components are medicine and health, as well as financial systems and educational systems.

THE SETTING

Synthesis is the process of making or producing what is otherwise natural, but also of making things that nature could not or did not make under natural conditions.

Analysis has dominated during the present 20th Century, and was built on the developments, often philosophical in nature, of its predecessors in previous centuries—ideology, enlightenment and reason. For the next century synthesis will continue to grow in importance as a result of the increasing involvement in conceiving, building, and assembling of artifacts and systems. This process of synthesis will itself be the subject of considerable analysis.

Synthesis of a thing that is already existing (a substitute, an imitation), such as synthetic rubber, is different than synthesizing something previously unknown, particularly a large system, such as a worldwide communications system, especially the original one. The first approach involves considerable analysis of the original material or system that may be used to inspire or guide the work; the second requires a creative approach into the less well known. Synthesis can provide a way of projecting thoughts to meet future needs and to satisfy wants. Synthesis provides a framework for guiding analysis, research, development, management and education.

There will be much interchange of terminology between philosophical and natural and unnatural or artificial systems. Terminology used in analysis and synthesis was first developed with respect to philosophical considerations, then was applied to natural systems, and now is being applied to artificial systems. Philosophers of old wrote of atoms without knowing what they were, and the term was used and is presently used extensively. Lavoisier (1743-1794) referred to " ...elements as the last point reached by analysis... " referring to chemicals and the state of knowledge in his day. Terminology developed for physical systems is now being applied to philosophical systems, particularly as related to energy and the environment. Philosophers use terms of scientists, such as relativity, time, and space.

BRIEF BACKGROUND

The Age of Synthesis is based on the premise that philosophical thought has developed and evolved from the earliest days of learning, usually rooted in the search for truth, undoubted by preceding written history. For example, early weapons began with the materials literally at hand and the desire to improve the capabilities of weapons became a driving force for new methods and materials. Developers responded to "what if....? " an approach still used, elevating synthesis as an effective approach. Although as we look back today we might say that the early thinking was primitive, those beginnings, as each step in the development provided a new foundation for the next step. It is probably only in the last 200 years that the objective of governments changed from weapons to weapons along with some other actions to benefit the population.

Philosophical thought was developed on concepts that became the basis of scientific approaches, categorized as natural philosophy. During the early years the term "philosophy" was generally abandoned for real world objects and applications. Those who embraced things or artifacts of the world were not considered philosophers. Early developments in philosophy consisted of developing the thought process and expressing the thought process in the language. Early religious and legal thought was built on synthesis of knowledge. Later and recent developments, consisted in applications to things, thus there has been an evolution over the centuries from thoughts to words to things. (Image 1-1)

'Analysis and synthesis' were among the terms used to describe the thought

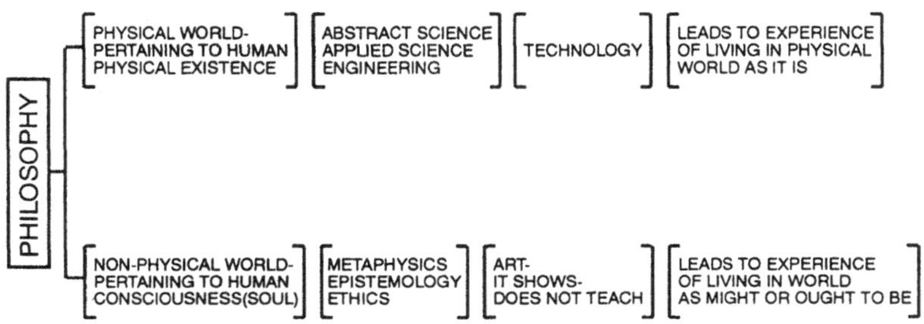

Image 1-1. Philosophy, physical and non-physical

processes. Except for a few people such as Archimedes, Michelangelo, Leonardo da Vinci, and Shakespeare, whose exploits of creative thinking and synthesis approaches are well-know, analysis was dominant and was the basis for most of the thinkers. 'Analysis and synthesis' have been recognized as methods of exercising the thought process to things resulting from thinking. As systems became more complex, particularly unnatural systems, there was an increase in the amount of synthesis involved.

Chemistry has used synthesis as a means of duplicating, substituting, and imitating as a major effort; biology has combined evolution and synthesis. Architecture and engineering, those professions with a design thrust as a basis of a built- or constructed-device or system as contrasted to a natural system, utilize synthesis to a much greater extent than scientists. The architects, in general, have recognized these relationships and have done a better job than the engineers in incorporating synthesis into the educational program, although they can be criticized for avoiding some of the basic subjects that would enhance the value of their synthesis.

Future developments of things and systems of things will incorporate to an increasing extent the synthesis approach—integration, interdisciplinarity, interconnectivity, imaginative, ..., holistic—and to larger and larger systems, not necessarily larger in size, but also in numbers of components and interactions.

Human systems (business, economics, education, government) will increasingly follow these patterns in the 21st Century, often combining human activities, behavioral systems, and artificial systems, in a mix of natural and unnatural systems. The wide use of computers has greatly increased the interactions between human, natural and unnatural systems.

CONCEPTS

The 'thoughts – to – things' process assumes that one initiates through thinking, then puts thoughts into a form that can be passed to others by speaking and writing, and then applies these thoughts to an imagined, natural, unnatural or artificial (built) thing or system. End use is a driving force in synthesis, whether innately as in a natural system or teleologically (design) in an artificial system. That flow is represented by:

THOUGHTS – to – THESES – to – THINGS

Thinking/Ideas	**Speaking/Writing**	**Something**
Abstract	Designing	Imaginary (dream)
Concrete	Sketching	Natural (discover)
Imaging	Publishing	Artificial (invent)

Analysis and synthesis are involved in each step from thoughts to things. It is probably impossible to identify an activity that is pure analysis or pure synthesis. Analysis predominates in the study and knowledge of natural things and systems. Synthesis predominates in development of unnatural or artificial (human made) things and systems. Words that describe the synthesis of the artificial include: systemization, interfaced, interconnected, integrated, interrelated, interdisciplinary and multidisciplinary, etc. (Matrix 1-1)

During the 20th Century, analysis has predominated in philosophy and science as people have learned much about themselves and nature. During the 21st Century, synthesis will dominate as the information of nature and the imagination of people are combined to produce an abundance of artifacts, artificial or unnatural systems. The products of synthesis will be the subject of considerable analysis, not only in a materials sense, but also in human systems. Likewise, there is an increasing recognition that there is a significant contribution to the creative process by the reverse flow from things to thoughts, that is, each new product (artifact) stimulates the creative process. As a result of the proliferation of artifacts and artificial or unnatural systems, the general public will become more involved in controlling, restricting, limiting these systems. Government will be empowered to play a major role, primarily in the control of unnatural systems, a reason for education to include synthesis in its toolbox of skills.

Matrix 1-1 Representations of Thoughts to Things

THOUGHTS to **THESES** to **THINGS**

Thinking
Imagining
Idea-ating

converted by

Speaking
Writing
Sketching
Drawing
Designing
Modeling

converted to

Devices
Assemblies
Systems
Sculpture
Painting
Literature
Policies

THOUGHTS – TO – THINGS

THOUGHTS— there is a synthesis of thoughts as a result of thinking within one's intellectual system. Thoughts are based on knowledge, experience and imagination, producing data, information, images and ideas. Locke says that there are three degrees of knowledge: intuitive, demonstrative, and sensitive that which can be sensed. Analysis is involved in this domain to sort out the information for the thesis or other transfer functions from thoughts-to-things. Adler (1977) originally published 'The 102 Ideas' of the Western world in 1952, an irreducible list that still stands. (Matrix 1-2) These ideas were a part of the Greek culture 2000 years before. Neither 'analysis' nor 'synthesis' appear on the list, but 'universal and particular' appears as an entry, illustrating the complementary nature of the analysis-synthesis.

THESES— there is a synthesis of ideas and thoughts from numerous people, from the written, spoken, and visual representations providing a design, publication, treatise, or speech. Analysis is involved in this domain to sort out and develop the pertinent information to help meet the end use of things. As thoughts are dispensed, activities can move from individual to group to team efforts.

THINGS— there is a synthesis of different things (artifacts) for replacement of items or systems, for a larger sphere of items or things, and for new items or systems. An action is required to accomplish THINGS. Analysis is

Matrix 1-2. The 102 Ideas

Angel	Family	Man	Quality
Animal	Fate	Mathematics	Quantity
Aristocracy	Form	Matter	Reasoning
Art	God	Mechanics	Relation
Astronomy	Good and Evil	Medicine	Religion
Beauty	Government	Memory and	Revolution
Being	Habit	Imagination	Rhetoric
Cause	Happiness	Metaphysics	Same and Other
Chance	History	Mind	Science
Change	Honor	Monarchy	Sense
Citizen	Hypothesis	Nature	Sign and Symbol
Constitution	Idea	Necessity and	Sin
Courage	Immortality	Contingency	Slavery
Custom and	Induction	Oligarchy	Soul
Convention	Infinity	One and Many	Space
Definition	Judgment	Opinion	State
Democracy	Justice	Opposition	Temperance
Desire	Knowledge	Philosophy	Theology
Dialectic	Labor	Physics	Time
Duty	Language	Pleasure	Truth
Education	Law	Poetry	Tyranny
Element	Liberty	Principle	**Universal and**
			Particular
Emotion	Life and	Progress	Virtue and Vice
Eternity	Death	Prophecy	War and Peace
Evolution	Logic	Prudence	Wealth
Experience	Love	Punishment	Will
		Quality	Wisdom
			World

Adler, Mortimer J. 1952. The Great Ideas-Syntopicon. Encyclopaedia Britannica, University of Chicago, Chicago, IL 2 vol

involved to provide the most appropriate artifact for the end use such as in theeconomy,energy, material, and environment. Aristotle speaks of 'productions' that could be considered as things. As one moves from left to right on the activity relationships, there is an increasing utilization of synthesis of THOUGHTS and THESES and THINGS. In the reverse direction there is an increasing use of analysis that is done to improve the results of synthesis.

The world is made up of many things, big things, even earth itself. The importance of growth and synthesis may be more fully appreciated if one contemplates an answer to the statement—name a thing, anything, that started big.

Thinking is the ability to draw inferences and solve problems and answer questions on the basis of information from perception and memory. (Meyers, 1987) Perception has a strong pull on the process. In many human processes, such as in politics and advertising, much effort is exerted to assure that perceptions of the person or product are favorable (or that the other person or product are unfavorable). In the sciences one may pursue and consider perceptions but search for the facts related to the perceptions and must work to match perceptions and information.

Synthesis is a way of THINKING, an attitude, an approach, a way of doing, followed by action to meet the end use objective, with an accomplishment or non-accomplishment, if not feasible, to produce THINGS. The synthesis approach adds something new to the subject and process, not necessarily a part of it.

SYNTHESIS INSPIRED BY THOUGHTS

Some people have the ability to inspire others to think, even though the listener does not comprehend the subject, by their challenging speaking or writing, their proposal of ideas, and their facility of expression. After listening to Samuel Taylor Coleridge (1772-1834) for two hours without interruption, William Wordsworth (1770-1850) said, "Pray, tell me did you precisely understand what he said...? " His friend Rogers said, "If the truth must out (sic), I did not understand a syllable from one end of his monologue to the other." (Lee. 1954)

R. Buckminster Fuller (1895-1983) could talk for three hours at a one hour lecture putting forth hundreds of ideas, stimulating listeners of various backgrounds who often did not understand much that was said. Isaac Asimov (1920-1992) was an imaginative and prolific writer who excelled in both fiction and non-fiction. He wrote on numerous subjects, and was a master at making the difficult more easy to understand, often by presenting the pieces as a part of the whole.

Aristotle said that "All 'productions' come to me from art or from capacity or from thought. Some of them, too, come to me spontaneously and by chance, in much the same way as do some of the things that come to me by nature. In these processes, one part is called 'thought' and the other 'production.'"

Although there are many benefits and new insights to be gained by using synthesis, there is a danger of being misled by myths and miracles. Thoughts evolving from myths, imagination, and perceived miracles become inspiration for new approaches but need to be challenged and tested. Control is needed to prevent synthesis from being used as a vehicle for promoting the occult or witchcraft.

It is difficult to capture the depth of understanding of Socrates, Plato, Aristotle, Kant, Emerson, and the others who we anoint for their insights on analysis and synthesis. They never dreamed that we would be applying their ideas to our THOUGHTS to THINGS of today, particularly to unnatural or artificial things or artifacts.

The 'thoughts – to – things' relationship as a part of a synthesis view is illustrated by:

> The Tao is something blurred and indistinct,
> How indistinct! How blurred!
> Yet within are images,
> How blurred! How indistinct!
> Yet within are things.
>
> *Chinese Philosopher Lao-tzu [born c. 600 BC] (Hughes,1991)*

SUMMARY

Ideas are the result of thinking. These ideas are the precursors of more definitive knowledge about natural things and systems. These ideas become the framework of knowledge needed to build unnatural or artificial things and systems, thus a flow of thoughts-to-things, with a transfer function between. The first action is more likely to be prominently analysis, the second, synthesis. An interplay between analysis and synthesis is involved, each helping the other along to some higher knowledge and understanding of natural and unnatural things and systems. Synthesis is a way of thinking as well as the procedure for making a thing. Synthesis involves thinking with a goal, an end, an objective in mind, involving integration of elements, components, and subsystems to form a system.

Synthesis has obviously had an important role in developing new unnatural or artificial products or artifacts—substitutions, imitations. Artificial systems are becoming larger, include more components, and are more complex than in previous systems. Synthesis thinking is becoming recognized throughout our society as an important approach for developing and improving human systems, many of which involve not only natural systems, but unnatural or artificial systems as well. Education should include a greater emphasis on synthesis, as government and the population in general becomes increasingly involved in controlling artificial systems.

Part 2.
Philosophy & Early Philosophic Thinking

INTRODUCTION

Philosophy was the earliest scholarly activity. Early philosophy dealt with the thinking process, and was involved in religion, ethics, government, arts, and law. Early philosophy dealt with the synthesis of knowledge. Early attempts to interpret or use philosophy for physical systems were discouraged, and could have delayed the development of philosophy. Archimedes' thinking was counter to the philosophers of the day because he tied his thinking to 'things' or physical systems.

Activities of scholars evolved into the sciences, primarily biological and natural, with the early study of sciences known as natural philosophy. Its counterpart dealing with the non-science subjects is moral philosophy. A distinct philosophy of engineering has not emerged but now appears to be emerging. The latent development of a philosophy of engineering can be attributed to the young age of the profession, because the roots of the profession lie in both the arts and the sciences, and because of the reluctance of philosophers to get too closely identified with use or application. The early years of the engineering profession were dominated by the crafts. Then later the profession was defined as "the art and science by which the properties of matter and the sources of power in nature are made useful to man in structures, machines and manufactured products." The first degree program in engineering at the U. S. Army Academy was identified as natural philosophy. The sciences came to the fore in the middle of the 19th Century. Building on the strengths and attributes of the arts and sciences, a unique engineering profession has evolved in practice and is still seeking to identify a unique education that emphasizes synthesis, particularly of artificial or built or constructed devices and systems. Analysis is an important aspect of engineering but synthesis becomes the overriding attitude and goal of engineering practice.

There are some who see the approach of the practice of the profession of engineering as closer to that of a poet than the scientist. The common goal of the poet and the engineer is to develop a concept, to prepare a design, and to make a product. A view of the poet is expressed by Emerson (1893) who writes about the limitations of analysis:

> " ... the analytic process is cold and berieving (sic). The poet sees the whole and avoids analysis."

AGES OF PHILOSOPHIC THINKING

The philosophic ages of the past to the present have been roughly categorized by centuries as follows:

1900s	XXth	Age of Analysis
1800s	XIXth	Age of Ideology
1700s	XVIIIth	Age of Enlightenment
1600s	XVIIth	Age of Reason
1500s	XVIth	Age of Renaissance

It is impossible to define precisely the beginning and ending of many events, eras, and evolutionary developments. Only with the benefit of historical perspective can these ages be determined, and even then it is difficult to gain a consensus on the categorization. Certainly the identification of ages by centuries is open to considerable question, as the identified ages extend into adjacent centuries, the beginnings or the roots of which may go back several centuries. (Matrix 2-1) A new age or era begins long before it is identified, in the bowels of the previous age, and then continues far into the next century but is overtaken by the newer age.

We tend to think that the use of electric devices for making music developed in the last 30 years. (See Part 5) However, the electronic (or electric) honors go to the Dynamophone built by Thaddeus Cahill between 1895 and 1911. Although this device filled two freight cars and weighed 200 tons with its own generator, it was the first to use electrical oscillations to control individual components of sound. And one can go back even further to 1759 when Jean-Baptist de Laborde's Clavecin Electrique was the first instrument to use electricity in the mechanical production of sound. A static charge stored in Leyden jars controlled by depressing a note on a keyboard, caused a clapper to oscillate between two bells tuned to that pitch. One can also go back even further, as early as 1624, Francis Bacon had already described modern music technology. (Fritts, L. 1993) Similar findings are found if one attempts to document the first automobile, the helicopter, or to define a physical law, or describe an atom.

The elements of the Age of Analysis really began in recorded history before the birth of Christ, with Socrates and his followers, but was used to identify the twentieth century because so many activities from religion to science were "analytic" in nature.

In the 21st Century the Age of Synthesis will build on the Age of Analysis of the 20th Century, philosophically and technologically. Synthesis builds on analysis just as the causal relationships by systematic experiment (Renaissance) built on the formal system of logic by the Greeks (Euclidean geometry). The Age of Synthesis took major root during the Renaissance, although there was evidence of synthesis thinking in BC by Archimedes. It is likely that synthesis

Matrix 2-1. The Ages of Philosophic Thinking

2000s	XXIst Century	**Age of Synthesis** (suggested)
1900s	XXth Century	**Age of Analysis** by Morton White Bergson, Croce, Dewey, James, Peirce, Russell, Santayana, Sartre, Whitehead, Wittgenstein.
1800s	XIXth Century	**Age of Ideology** by Henry Aiken Compte, Fichte, Hegel, Kant, Kierkegaard, Mach, Marx, Mill, Nietzsche, Schopenhauer, Spencer.
1700s	XVIIIth Century	**Age of Enlightenment** by Isaiah Berlin, Berkeley, Condillac, Hume, Locke, Reid, Voltaire.
1600s	XVIIth Century	**Age of Reason** by Stuart Hampshire Bacon, Descartes, Galileo, Hobbes, Leibniz, Pascal, Spinoza.
1500s	XVIth Century	**Age of Adventure—The Renaissance Philosophers** by Giorgio de Santillana Bruno, Copernicus, Leonardo Da Vinci, Erasmus, Galileo, Kepler, Machiavelli, Michelangelo, More, Montaigne.
1500-1400s	XVth-XVth Century	**The Age of Belief—The Medieval Philosophers** by Anne Fremantle Abelard, Anselm, Averroes, Boethius, Bonaventura, Erigena, St. Augustine, St. Thomas Aquinas.

predated analysis as a method of approach preceding the time of the Christian era, particularly in the Middle East, Roman, Egyptian, and Greek empires. The noticeable beginning of the present technological synthesis could be argued to have begun in the 1930s with the substitution of plant materials for metals and oil, at that time identified as chemury. Chemury was popularized by Henry Ford's use of the soybean for automobile parts and Dupont's nylon for hose. From then on substitution of manufactured or processed organic materials for some naturally occurring substances began in many fields as availability, characteristics, and price of certain natural resources became limiting. These actions demonstrated that analysis and synthesis were important economical approaches not only for substitution of materials but for new applications.

Socrates, Plato and Aristotle provided early extensive philosophical

Socrates 470 BC-399BC **Plato c.427BC-c.347BC**

Aristotle 384BC-322BC
Magnus, A. 1193-1280 Follower of Aristotle
Aquinas, St. Thomas 1225-1274 Student of Magnus; Aristotelian
Leonardo da Vinci 1452-1519 Synthesist
Copernicus 1473-1543 Experimentalist
Michelangelo 1475-1564 Synthesist
Galileo 1564-1642 Experimentalist
Bacon, F. 1561-1626 Inductivist; Experimentalist
Kepler, J. 1571-1630 Mechanist
Hobbes, T. 1588-1679 Mechanist; Materialist
Leibniz, G. 1646-1716 Aristotelian
Descartes, R. 1596-1650 Mechanist; Speculative thought
Locke, J. 1632-1704 Empiricist
Newton, I. 1642-1727 Mechanist
Berkeley, G. 1685-1753 Idealist (religious)
Hume, D. 1711-1776 Empiricist; Skepticist; Cause and effect
Kant, I. 1724-1804 Rationalist; Transcendentalist, Idealist

Fichte, J. 1762-1814
Subjective idealist
Hegel, G. 1770-1831
Dialectic; Synthesist
Comte, A. 1798-1857
Positivist
Emerson, R. 1803-1882
Transcendentalist
Mill, J. 1806-1873
Positivist; Empiricist
Spencer, H. 1820-1903
Positivist
James, W. 1842-1910
Radical empiricist
Eucken, R. 1846-1926
Idealist
Husserl, E. 1859-1938
Phenomenologist; Existentialist
Dewey, J. 1859-1952
Pragmatist; Instrumentalist
Whitehead, A. 1861-1947
Logical positivist
Russell, B. 1872-1970
Logical positivist
Moore, G. 1873-1958
Intuitivist (ethical); New realism
Einstein, A. 1879-1955
Intuitive synthesist; Stringent empiricist
Carnap, R. 1891-1970
Logical positivist; Analytic phil.
Wittgenstein, L. 1889-1951
Logical positivist
Heidegger, M. 1899-1976
Logical positivist; Existentialist
Sartre, J.P. 1905-
Existentialist

Image 2-1. A Succession of Philosophers

Image 2-2. People Who were Most Effective in Defining Analysis and Synthesis

approaches. These people formed the basis for a rich philosophical tradition in the Western world. Undoubtedly there were many other individuals, but who for lack of publications or preservation of manuscripts, have been deprived the recognition for similar contributions.

A person of the ancient period (before 470 AD) who could be considered as using a synthesis approach in the physical world, based on written history, was Archimedes (287-212 BC). Others who followed were Roger Bacon, Francis Bacon, Leonardo da Vinci, and Michelangelo. Kant wrote extensively on analytic and synthetic statements from a philosophical view. Many people, such as Herbert Spencer, built on the work of Kant. (Image 2-3) Modern day synthesists in philosophy include Husserl, Emerson, Sartre, Teilhard, and Tillich and in science and technology include Steinmitz, Edison, Bohr, Kettering, and Fuller. Bohr (1885-1962) synthesized ideas when in 1926 he synthesized the two separate long-standing ideas that light is both a wave and a particle, thus advancing a theory of complementarity.

Rationalism
Rene Descartes (1596-1650) (Platonic School)
Gottfried Leibniz (1646-1716) (Aristotelian School)

Empiricism
John Locke (1632-1704)(Platonic School-on/off)
George Berkeley (1685-1753) (Platonic School)
David Hume (1711-1776)

A Synthesis of Rationalism and Empiricism
Critical Idealism **Transcendental Logic**
Immanuel Kant (1724-1804)

Neo-Kantianism, from 1870-1920

Marburg School	Hermann Cohen (1842-1918) Paul Natorp (1854-1924)
Physiological School	Hermann von Helmholz (1821-1894) Friedrich Lange (1828-1875)
Realist School	Alois Riehl (1844-1924)
Heidelberg or Baden School	Wilhelm Windelband (1848-1915) Heinrich Rickert (1863-1936)
Gottingen School	Leonard Nelson (1882-1927)
Realivist School	George Simmel (1853-1918)

Based on Flew (1984) and Runes (1968)

Image 2-3. Kant's Predecessors and Followers

Building on the philosophy and experiences of the past, even as analysis is strengthened and improved, synthesis will be more important to help direct the resources and talents used in analysis to meet the needs of society. The synthesis approach, by helping to identify gaps in knowledge, will help define the needed research and development areas which might best be carried out by analysis. Synthesis needs to be more consciously recognized, taught, utilized, and experienced. As the concerns of society are impacted by an increasing number of unnatural systems, and as people abuse natural systems, members of society should be better taught to assure more intelligent decisions. And to teach synthesis one must experience it. The building blocks of synthesis need to be identified more clearly. Tentatively, it would appear that the building blocks of synthesis include:

» considering factors outside of existing identified or recognized natural and unnatural systems

» looking to a larger system as the framework of reference

» visualizing the results of the activity

» linking and interacting of parts

» involving people of various interests and backgrounds in the process

A potential weakness of this approach is that too many people might opt for synthesis without recognizing the importance of critical analysis.

EARLY USERS OF SYNTHESIS

Early civilization took root from the efforts of those who used synthesis. Beginning five hundred years before the birth of Christ, the Egyptians built the famous pyramids, the only remaining artifacts of the Seven Wonders of the Ancient World. These and the irrigation systems, water and waste systems, the canals, temples, fortifications, and many more built structures throughout the Inca Empire, Egypt, Roman Empire, and the Middle East were the products of synthesis. People had to perceive, imagine, and construct the devices and systems without scientific knowledge as we know it. Many of these products of synthesis, in fact, laid the basis and demonstrated the need for analytical science to develop. These developments laid the foundation for science throughout the world, contrary to the often perceived birth of science by Europeans.

Archimedes (287?-212 BC) is one of the best known early inventors who developed water handling devices, was also an outstanding mathematician and renown in mechanics. He began to use and develop scientific principles and synthesized knowledge and experience. He was known for his mechanical inventions as well as his mathematics prowess, and was condemned by Plutarch who said of himself that he

> "would not design to leave behind him any commentary or writing or such subjects since he considered them sordid and ignoble"
> *From Great Books of the Western World, vol. 11, p. 399*

Plutarch wrote, " ...repudiating as sordid and ignoble the whole trade of engineering, and every sort of art that lends itself to mere use and profit... " (deCamp, 1974)

Similar thinking was still prevalent with some scholars in the 20th Century who disdained those who worked in subject areas where the results had practical applications.

Socrates (470-399 BC) and Plato (427-347 BC) and Aristotle (384-322 BC) came along later and could have used the artifacts of the time to accompany their philosophical development, but they did not. In fact, "many ancient writers, such as Cicero, like Plutarch, express the idea that the essence of a true gentleman is his refusal to have anything to do with things of practical utility." (deCamp, 1974). That attitude undoubtedly discouraged many thinkers of the time who then avoided association with the practical activities, and delayed the further development of analysis and synthesis as working tools to meet the needs of potential users.

During the thousand years from the time of Socrates until the Renaissance few people in Europe made notable advancements in the scientific area. The Chinese were very busy in this period but went little further, perhaps because of a stalemate on inquiry. In the scientific area, one of those who made advancements was Roger Bacon (1214?-1294), who might be considered a synthesist, who prepared the encyclopedic Opus Magus that incorporated treatises on logic, grammar, mathematics, physics, philosophy, and philology, and experimented with the development of gunpowder and optics. Bacon's writings included speculations and imaginations of artifacts that could be built, generally considered as synthesis in approach, even though the items were not built, providing precursors of things that could come in the future.

The beginning of the Renaissance was taking shape in the 15th Century. As respect for learning returned, more people became immersed in educational activities, governments began to spend money for learning, arts, and civil projects, and individual accomplishments were again recognized.

The Renaissance, or the 16th Century, marked a return to learning that had languished for centuries. The best known of that period were Leonardo da Vinci (1452-1519), painter, sculptor, engineer, and architect, and Michelangelo (1475-1564), sculptor, architect, artist, and poet. Leonardo da Vinci would probably never have divided art and science into two fields. Both Michelangelo and Leonardo were great visualizers of the possible as represented by their work and notebooks, and would have to be included in a list of world's great utilizers of the synthesis. When Leonardo da Vinci applied for work at the Vatican the Pope said of him, "he always thinks of the end before the beginning. He'll not amount to anything." He didn't get the job. The talents of synthesists are often overlooked and not appreciated. Covey (1991) advises people "to begin with the end in mind" if they are to be productive, a trait of a synthesist.

The roots of the 'modern' age began to grow as the authority of the church was challenged, a merchant class developed, and navigation supported trade. Others associated with the Renaissance are Copernicus (1473-1543), Galileo (1564-1642), and Kepler (1571-1630) and Newton (1642-1727), who was born in the year of Galileo's death. These people synthesized available information to postulate on physical phenomenon which in turn was analyzed by many others for years to come. These people caused a paradigm shift in thinking following their work in the 16th and 17th Centuries. (Kuhn, T. 1962)

PHILOSOPHIC THINKING PRECEDED

The speaking and writing about the philosophies of thinking and thought, often with concepts that are unprovable, precede the use and application of that thinking. From a theological point of view "the gods are normally synthesized as so many manifestations of the one being which have been adapted to the level and situations of the adherents." (Edwards, 1972) The thinking in the early years about religion and humans became a basis for thinking about biology, physics, and mathematics. There is a transfusion of the thoughts of "pure" moral philosophy to natural philosophy, and the associated terminology gets projected to such applications. The background or field of interest of the early philosophers shows in their thinking as they espouse their philosophy, which points to the difficulty of maintaining a pure philosophy similar to the difficulty that some people have as they struggle to maintain pure science. Thus, Herbert Spencer (1820-1903) writes in terms of biological organisms: George Boole (1815-1864), in algebra; Rudolf Carnap (1891-1970), logic in mathematics; Antoine Lavoisier (1743-1794), in chemistry; and Isaac Newton (1642-1727), in physics. Most of the philosophers of the last four centuries studied in mathematics and science subjects, particularly physics, building on these subjects to become productive in philosophy.

Beginning with Socrates, and his followers, Plato in generalization and Aristotle in analysis, their thoughts and the theses developed by them have had a tremendous impact on the philosophies that followed. Most of the leading philosophers were not synthesists, but the work of the philosophers laid a foundation for the synthesis approaches now coming into vogue. There was about a two hundred year gap in a strong synthesis approach between philosophers in the Age of Renaissance and philosophers in the Age of Ideology. During that period mathematics and the sciences emerged as major activities providing a firm base for analysis, and with analysis emerged synthesis.

The feedback from science and technology to philosophy influenced the development of positivism, dualism, materialism, pragmatism, mechanisticism, instrumentalism, and phenomenology. An example of the impact of science on philosophy is shown by dualism (Matrix 2-2), a view that in some areas of concern there are two fundamental entities, such as the view in metaphysics that a person's mind and body are two irreducibly different entities; the view in religion that good and evil are in opposition; the view that there are true and false statements; and in electrical physics that there are positive and negative charges. Analysis and synthesis can be considered as an approach in dualism, and can be applicable to arts, sciences, applied science fields, and professions.

Matrix 2-2. Philosophers Who Contributed to Dualism

Zoroaster (c. 660 B.C.), who lived in Persia had a strong element of dualism, good versus evil, in his mystic thought. He is often given credit as being the first philosopher to establish good and evil as abstract values and well as for individual behavior.

Plato (428-348 B.C.) is famous for his thoughts on the dualism of mind and body.

Descartes, Rene (1596-1650) was a rationalist who developed the 'method of doubt' philosophy, and that a human being is some kind of a union of soul and body. He used concepts in dualism as he developed important concepts in geometry and is credited with starting the cartesian approach.

Kant, Immanuel (1724-1804) is associated with the philosophy of transcendental idealism who maintained that knowledge can only be synthetic and a relative product of logical consequences. He moved the thinking as related to duality out of metaphysics into epistemology that became the contrast between formal and material, transcendental and empirical, and analytic and synthetic elements.

Hegel, Georg W. F. (1770-1831) is characterized as an objective idealist. His approach is characterized by the use of dialectic thesis, antithesis and synthesis, maintaining that the world develops dialectically by the interplay of opposites, a representation of duality.

Analytic and synthetic thinking are in a way parallel to yin (passive) and yang (active) in the behavioral field, deductive and inductive thinking in the sciences, and warp (long cord) and woof (short cord) in fabric as representative of the technical world.

DUALISM

The evolution of thinking in dualism can be seen from the contributions of philosophers listed in Matrix 2-2. Dualism should be considered in terms of two parts of the whole with interaction between the two parts, and not as a bifurcated phenomena. Bifurcation implies that two parts develop separately without interaction, except at the top or beginning of the process or method.

Descartes (1596-1650) believes that the method of synthesis is not as satisfying nor the learning mind as engaged as in the method of analysis. Synthesis employs a long series of definitions, postulates, theorems, axioms, and if a challenger denies one of the conclusions, it can be shown that the overall conclusion is not acceptable. With analysis the parts are learned and more likely accepted by others. Hopefully the recognition of the need for more synthesis, as demonstrated by potential users needs, and a more broadly educated public, synthesis will be as challenging as analysis. Hobbes (1588-1679) 'second order' of ideas says something general about a whole class of 'first order' statements, a relationship of synthesis and analysis. Hobbes was more of a synthesist than analyst.

SUMMARY

Early philosophy dealt with thinking, void of experimentation involving 'things.' There was a disdain for relating philosophy to things of the real world. That disdain may have delayed the development of a comprehensive philosophy in which there would be stronger ties to other fields. The early 'thoughts' of philosophers were devoted to religion, law, and governance. These efforts were usually synthetic in nature. Many of the philosophers came from a background of mathematics and physics. The next step was to consider the 'things' of the natural world in physics, biology, and chemistry. These efforts were primarily analytic in nature to better understand the natural products and processes. The next step that is being followed, and will develop into a greater maturity by considering, in addition to the natural, the unnatural or artificial world of the scientist, medical doctor, engineer, agriculturalist, and business fields. Synthesis will come back into the picture as a major method. These fields have a special role to fulfill in society by considering in the process of creating the artificial to put major emphasis on 'the end' or the 'user' of the synthesis. The dualism of analysis and synthesis will more likely meld into one activity, utilizing both methods of approach to assure that the optimum results will be obtained for society.

Part 3.
Methods of Thinking

INTRODUCTION

The methods of thinking, reasoning, and dialectical argument are divided into two major categories—analytic and synthetic. Neither category is completely independent of the other. Good analysis involves synthesis, and synthesis involves analysis. Refinements to analytical thinking include such approaches as deductionism, reductionism, and logicism. Refinements to synthetic thinking include such approaches as inductional, transcendental, existential, and phenomenological. Analysis provides a partial approach to evaluating synthesis, while synthesis provides a partial approach to identifying subjects for analysis. These approaches were first developed as a means of searching for truth related to thinking and reasoning. These approaches were then applied to biological and natural physical systems. These principles moved from a moral (ethical) system by the early philosophers, to living systems or natural systems by the philosophers-scientists. Only in the last 100 years have artificial systems attracted the scholarly attention of philosophers.

Intuition, perhaps tuned by previous experiences, plays a major role in the advancement of thinking, but is unpredictable. Johann Kepler (1571-1630) stated the second law of planetary motion based on intuition. The second law is that an imaginary line drawn from the sun to a planet moves over equal areas in equal times. Sir Issac Newton stated the binomial theorem by intuition theorem without proving it. The binomial theorem is an algebraic expression that gives the result of multiplying a binomial by itself any number of times. Intuition can help advance understanding. Incorrect intuition is not long concealed, and in the process learning occurs.

Intuition probably can't be taught, but it is known that the environment is an important variable for promoting or discouraging intellectual activity.

Philosophy is a search for truth through logical reasoning, rather than by quantitative observation. Some philosophers maintain that as soon as a relationship can be quantified it is no longer in the philosopher's domain. Bertrand Russell observed, "Every new scientific finding, quantifiable, robs philosophy of areas of study."

A person doesn't take pen in hand and say that today I am going to consider this problem by analysis, and tomorrow by synthesis. Each person develops an approach to problem-solving, usually all people utilize both analysis and synthesis, one approach to a greater extent than the other, depending on the person. Problem-solving usually begins by taking an overall view of the problem before getting involved in the details (synthesis). Those who get involved in the details too early in the problem-solving activity may lose sight of the overall objective (analysis). An artful combination of the use of analysis and synthesis will help solve the problem.

Thinking (or reasoning) can be by induction and/or deduction. Inductive thinking and deductive thinking apply to both analysis and synthesis. Intuitive thinking, although not considered scientific by many, may lead to unpredictable thoughts and things. John Stuart Mill's expressed his Canon of Rules on Induction (1843) as:

> " Empirical observation
> Careful gathering of facts
> Studying problems empirically
> Introducing experimental methods
> Correlating of facts
> Developing crucial tests."

Thinking can be by syllogisms. A syllogism is a valid argument having two premises, a major premise and a minor premise, and a conclusion that follows the arguments. (Image 3-1) Thinking can be by the dialectical process. The dialectical process provides for discussions between two opposed views with discussants accepting some of the views of the other until a final view, a result based on synthesis is reached. 'Groupware,' computer software, offers a technological tool (See Part 7) for using the dialectical process. (Image 3-2)

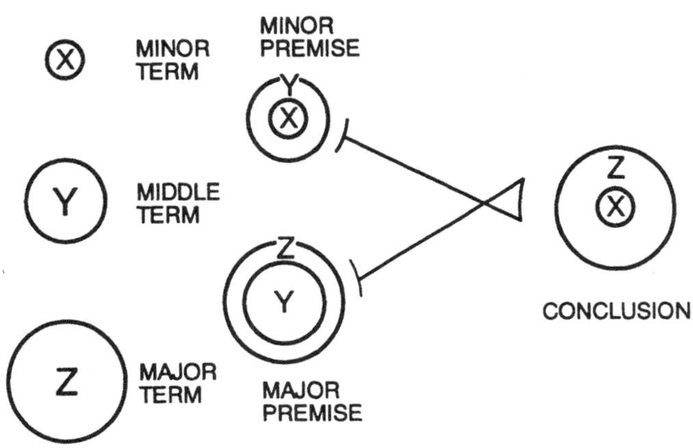

Image 3-1. Thinking by Syllogism (Brumbaugh, 1917)

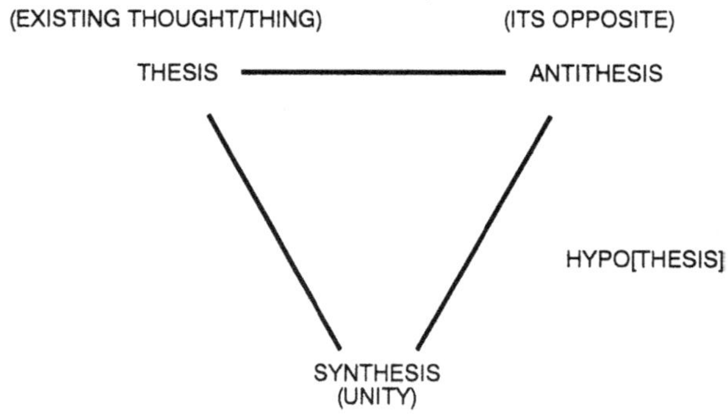

Image 3-2. Thinking by the Dialectical Process

ANALYSIS HAS DOMINATED IN THE 20TH CENTURY

Although analysis has dominated in the past, and will continue to be important, beginning sometime in the 20th Century, synthesis has increased rapidly and will predominate as an approach in the 21st Century. Fields of study, such as chemistry, that have been based predominately in analysis will continue but will have strong, productive components in synthesis, as evidenced by the expanding interest and work in synthetic chemistry and related subjects. Chemistry and biology that have been based primarily in analysis could be more appropriately identified as analysis/synthesis. Major breakthroughs in many fields of endeavor have resulted from synthesis such as Galileo in astronomy to the recent findings of Crick and Watson in DNA. A major breakthrough based on synthesis is more likely to cause a paradigm shift in thinking of others than with analysis, as witnessed from the DNA structural identification. Often human and political systems and organizational structures are formulated using a synthesis approach, with adjustments made as needed based on analysis. People in architecture, engineering, medicine, and other professions will recognize to a greater extent than scientists, the importance of synthesis in their activities, be distinguished from the science fields on which they depend, and be identified as synthesis/analysis. The synthesis/analysis context provides different approaches to problem-identification and problem-solving, and suggests how the first approaches are made to problem-solving. A scientist more often starts with analysis (looking inward); an engineer, looking at a humanmade system either imagined or in operation, more often begins with synthesis (looking outward). (Image 3-3)

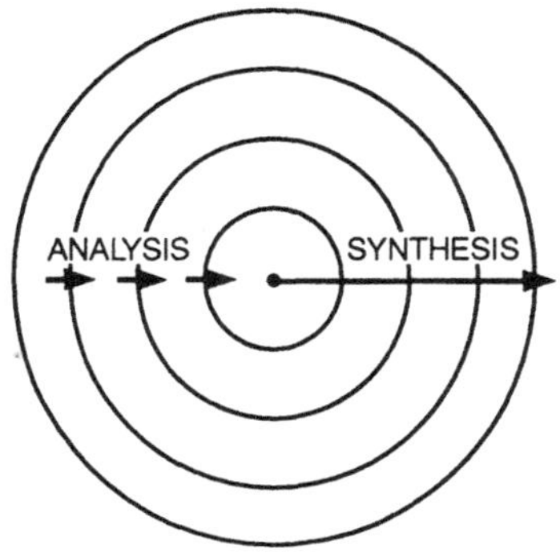

SYNTHESIS - PUT TOGETHER PARTS FOR A LARGER SYSTEM
LOOK FOR CONNECTIONS OF ONE PART TO
OTHER PARTS AND HOW THESE PARTS RELATE

Image 3-3. Analysis-Synthesis, The Same Whole

Analysis depends heavily on deductive thinking; synthesis includes inductive thinking while creative thinking involves both. Analysis is based on precision, such as 0.1, 0.11, 0.111, etc., and synthesis is based on an order of magnitude,

Matrix 3-1. Backward-Forward Looking — Toward An Integrated System

Backward Looking (attitude of analysis)	**Forward Looking (attitude of synthesis)**
Casuistic (based on data of past)	
Historians, judges, ministers, scientists, accountants	Applied scientists, professions (architects, engineers, lawyers)
Look backward to laws, rules, precedents as basis of thinking and decision-making	Look forward to anticipate consequences and base thinking and decisions on these

Looking backward and forward should be a part of the same integrated system. Obviously few people think and plan entirely as forward looking or backward looking. The above represents the emphasis on backward and forward looking (Hall, A. D. 1989)

10n. "Science can proceed only when it can observe with precision." (Bush, 1967) The new instruments available in the last half of the twentieth century have helped to revolutionize science by "seeing" things with precision and accuracy heretofore unseen or unknown. (Images 3-4 and 3-5)

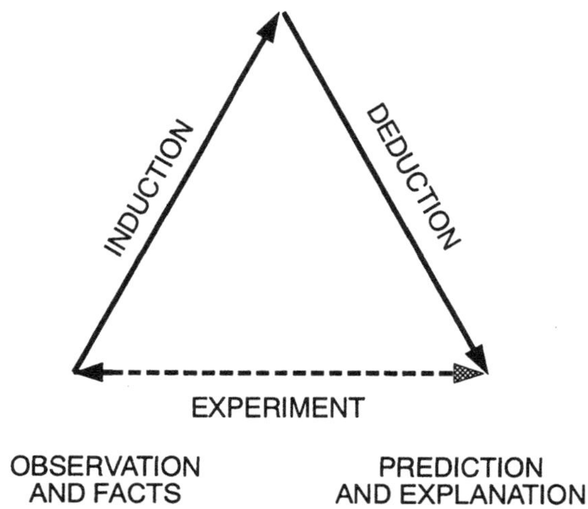

Image 3-4. Induction-Deduction-Experiment (Casti, 1990)

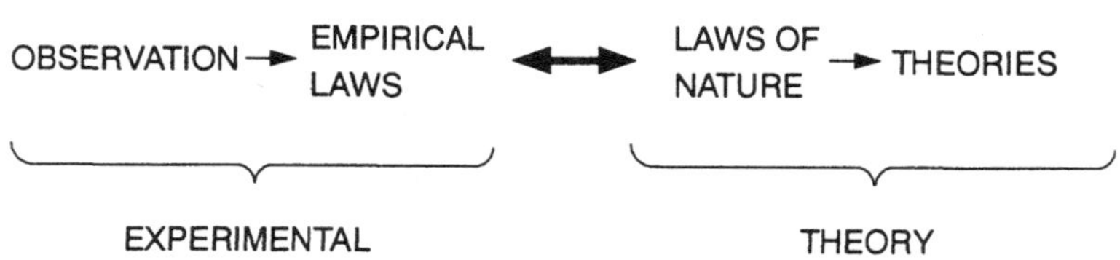

Image 3-5. Experiment and Theory (Casti, 1990)

Experts are generally considered to be analysts, but could be synthesists, and more likely will be so in the future. Many analysts are sympathetic to or understand the role and importance of synthesis. Whitehead wrote, "the more expert they were the less they would think of searching the concept of how society might enjoy the fruits of their discoveries." Fuller (Wagschal and Kahn, 1979) said " ...everybody today tends to believe that specialization is the best way to earn a living... As a consequence of comprehensively undertaken specialization we have today a general lack of comprehensive thinking... The concept of inevitable specialization by the brightest has become approximately absolute in today's social-economic reflexing. The fixation is false and is soon to be altered."

Most business, government, and educational organizations have employee jobs described and identified as analysts. In the 21st Century organizations will need employees who are integrators and facilitators and identified as synthesists or integrators. Problem-identification and problem-solving will consider to a greater extent the larger system in which an issue or question is imbedded.

Some fields of study and activities are dominated by people who think synthetically, think of a larger system, and consider the whole, although it may be difficult to identify the true whole. This includes poets, architects, artists, economists, and to an increasing extent, engineers who are challenged by creative thinking associated with artifacts.

There is a limit to the capacity of a person(s) to consider the whole, as many systems can be defined as made up of an infinite number of parts, layers, or hierarchies. The challenge is to think of a manageable system larger than what is envisioned at first consideration. The tools now available for instrumentation and computation make possible the consideration of larger systems than in the past and help accelerate the synthesis approach in many fields of endeavor.

It is interesting, and perhaps unfair, that poetry, literature, theater, writing, and artwork, all the result of synthesis, are practically always evaluated by analysts. But as someone (I believe it was Mark Twain) has said, "They haven't built any monuments to critics (often analysts)."

To accept the argument that synthesis will become more important in the coming century, one must also acknowledge that some philosophers do not accept synthesis based on inductive thinking. Peter Medawar believes that 'inductive and intuitive thinking as presented by Mills is not scientific and is a scientific fraud.' To overcome such criticism, analysis should be considered the major method of the sciences, and synthesis should be considered the major method of the arts, the professions and applied areas of engineering, architecture, medicine, agriculture, and education.

The variation between the thinking of people living in different countries is probably related to the thought leaders be they in religion, government, arts, sciences, and professions.

"In recent times continental philosophy has tended to be synthetic and British philosophy to be analytic." (Urmson, 1960). Conant (1964) characterizes the Germans as theoretical-deductive, and Americans as empirical-inductive.

APPROACH TO ANALYSIS/SYNTHESIS

On a practical level, synonyms illustrate a representation of the contrasts between analysis and synthesis, and are listed in Matrix 3-2, examples of which are: Analysis: synthesis; deductive: inductive; infer: intuitive; discover: invent; criticize: speculate; specialize: generalize.

Matrix 3-2. Synonyms for Analysis-Synthesis

As gleaned from writings by and about various philosophers

ANALYSIS/ANALYTIC	SYNTHESIS/SYNTHETIC
a priori, independent of experience	a *posteriori*, dependent on experience
Apollonian (detached, analytical)	Dionysian (involved, intuitive)
Aristotle	Plato (a generalizer)
Bodies extended	Bodies in relation to each other
Components	Systems
Convergent thinking	Divergent thinking (Hall, 1989)
Critical (concentrate on parts)	Speculative
Deconstruction	Construction
Deductive	Inductive
Differential	Integral
Disassembler	Assembler (Drexler, 1986)
Discover	Create
Divide, division	Sum, summation
Evolutionary	Revolutionary
Finite	Infinite
Fission	Fusion
Fractionate	Amalgamate
Idealism, idealistic	Realism, realistic (Hegel; Kant)
Idealistic	Pragmatic
Implicit	Explicit (Hegel)
Impressionistic	Expressionistic (Dudley, 1978)
Individual	Group, team
Inner consciousness (res cognitans)	Outer consciousness (res extensa)
Left brain (quantitative)	Right brain (creative)
Local	World, universe
Logic, logical	Nonlogical, empirical
Natural	Artificial, fabricated, manufactured, constructed, built
Objective	Subjective
Occident	Orient (Tagore, in Edwards)
Partial	Complete
Parts	Whole, gestalt

Predetermined thinking	Divergent thinking
Rational	Empirical, experimental
Reductionism	Agglomerate, amalgamate, assemble, blend, combine, formation, holistic, integrate, unite
Reductionist	Integrationist, expansionist
Representational	Romanticism (Dudley, 1978)
Rights (inner)	Responsibilities (outer) (Theobold)
Separate	Concrescence (Whitehead)
Separate, separation	Unify, unifying, unification (Hegel)
Simple element(s) to more	Complex elements
Specialist	Generalist, comprehensivist
Take apart	Put together
Theoretical-deduction	Empirical-induction (Conant, 1964)
Transcendental analytic	Transcendental dialectic (Kant)
Tuitive	Intuitive, intuition (Emerson)
	Existential (Sartre)
	Gestalt (sum of parts)
Vertical	Horizontal, lateral

All scientists are not analysts and all artists, engineers, etc. are not synthesists. The artist, engineer, applied scientist, poet, etc, usually start the thinking process with a view or vision of the item to be painted, written, made, assembled, built or manufactured to meet a want, need or imagination—that's synthesis. The succeeding steps involve assembling or inventing the appropriate materials, paints, and parts. That is followed by analysis, the results of which are fed back into the synthesis process.

The scientist will more than likely start with analysis, taking a thought, idea, or physical (natural or artificial) or biological body apart. The notable exception are the applied physicists, along with the engineers, as represented by their leadership in large projects requiring a vision in synthesis such as space exploration, ocean exploration, weapons, weather, accelerators and nuclear energy. Synthesis becomes more predominant when a society has 'discovered' nearly all there is to discover, and discovery goes on with analytic processes as improved tools are available for measurement and evaluation.

Synthesis is predominant in some fields of science, such as synthetic chemistry, in which a goal of a new or modified chemical is envisioned, followed by work to meet that vision. Once conceived, when a chemical is to be fabricated, considerable analysis may be used to learn how the chemical is put together, but is done within the framework of synthesis. A recent example is the discovery and development of C_{60}, buckminsterfullerene, or buckyball, that has stimulated a whole new approach to thinking about the structure of materials and research for properties, potential uses, and creativity. The fullerenes provide a new class of molecules that have potential use for high-temperature superconducting materials, molecular scale machines, superstrong electric transmission wires, and industrial catalysts. Breakthroughs in thought and developments that cause a step function in results are usually from the efforts of people with a vision that synthesize knowledge and experience and insight. Even those who 'accidentally' find a new phenomena or relationship or material, such as nylon or teflon, are exercising synthesis or the 'accident' would have been overlooked as others may have been overlooked. Teflon (polytetrafluoroethylene) was discovered by Roy J. Plunkett in 1938, and the monolith surface was soon used as a lining for cooking utensils, although that wasn't the objective. Many such accidents are not entirely found by accident, because the objective was to develop a new material, the work was being done in a laboratory by a prepared person who was expected to find something, which was done.

The recognition of the role of synthesis in science as the basis for major 'discoveries' about nature is becoming more prevalent. Synthesis is beginning to augment analysis, and perhaps gain top priority over the previous great age. There are now those who put greater value on synthesis than analysis as Covey (1989), "Interdependence [represents synthesis] is a higher value than independence," as he wrote of human systems.

EARLY RECOGNITION OF SYNTHESIS AS AN APPROACH TO THINKING AND REASONING AND MAKING JUDGMENTS

Immanuel Kant was the first to introduce the terms analytic and synthetic in philosophical discussions, and he is the most widely quoted person on the subject. Immanuel Kant (1724-1804) is widely quoted and referenced for his discussions of analytic and synthetic judgments in the book, "The Critique of Pure Reason." Another of his books, "The Critique of Practical Reason", although less well known, is equally valuable for studying analytic and synthetic relationships. Others, Leibniz (1646-1716), Hume (1711-1776), Spencer (1889. 1897) and Huxley (1942) added much to the literature on synthesis.

Although the pertinence of Kant's arguments and the accuracy of his assumptions for judgments are questioned by Runes (1968), the concepts and words have relevance for our discussion on the analytic/synthetic approach to thoughts, theses and things, even if not suitable for judgment of thoughts.

> "Kant raised the problem of judgment. Starting with the proposition that there are analytic and synthetic judgments, and laying down the rule that, by their *a priori* nature, apodicticity [to show, demonstrate] and universality are the criteria for scientific proof, he put himself to the question: How are synthetic *a priori* judgments possible? These judgments may not be derived from our knowledge, or our experience. This question of synthetic *a priori* judgment is the seed from which springs the whole 'Critique of Pure Reason.'

But the fact is that there are no synthetic judgments, nor any *a priori* ones—much less synthetic *a priori* ones." (Runes, 1968)

Kant distinguished between analytic and synthetic judgments. " ...an analytic judgment is one in which the concept of the predicate is contained in the concept of the subject, which in a synthetic judgment the concept of the predicate stands outside the concept of the subject." (Edwards, 1972)

A further distinction is made between *a priori* judgment and *a posteriori* judgment, first written about by Aristotle. *A priori* means provided without experience, that is, we know the truth without experience; *a posteriori* means knowledge dependent on experience, or insights into essential relationships within the empirically given. (See Appendix, Glossary). The accepted classifications of judgments are in Matrix 3-3.

Matrix 3-3. Classification of Judgments/Statements

	a priori	*a posteriori*
Analytic	yes	no
Synthetic	yes*	yes

*generally not accepted by empiricists

If analytic *a posteriori* and synthetic *a priori* are not generally accepted, analytic statements are *a priori* and synthetic statements are *a posteriori*.

Although Kant used these concepts to describe judgments, these concepts are useful for considering theses and things, be they natural or unnatural (artificial, artifacts). Carnap (1958) provides a visual representation of the true-false relationships of analytic and synthetic statements which can be written to include 'things' as well as statements. (Image 3-6) The use of truth-value tables (truth tables), attributed to C. S. Peirce, provides additional understanding as related to the truth and falsity of a propositional function under two or more basic propositions.

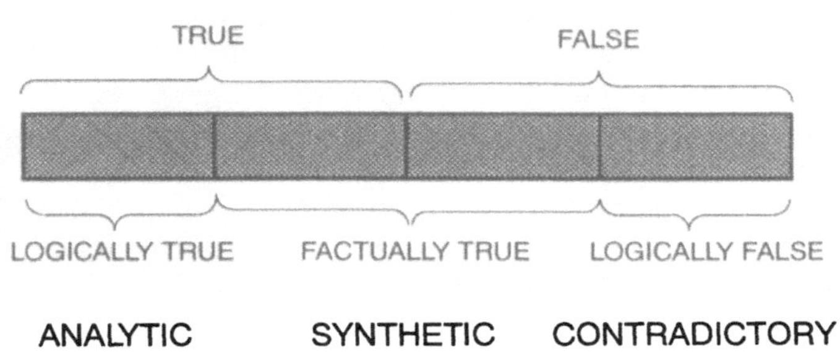

Image 3-6. True-False of Analytic-Synthetic Statements (Carnap,1958)

A PRIORI AND *A POSTERIORI*

These terms apply primarily to concepts, ideas, propositions, statements, or judgments. *A priori* means 'prior to experience' or prior to experimentation. *A posteriori* refers to being 'based on experience.' This term has been replaced in current literature by the word 'empirical.' (Lacey, p.10) Experience and empirical roughly mean that the senses are used to represent the concept or proposition. *A priori* concepts are sometimes call innate ideas. When a proposition is not known *a priori* , or not gained by experience (*a posteriori*), it is postulated.

Philosophers do not agree on the extent to which analytic and synthetic statements are *a priori* , or *a posteriori*. The major controversy is whether there are any synthetic *a priori* propositions. There is agreement that there are no analytic *a posteriori* propositions. There is agreement that the other two combinations exist; analytic *a priori* , and synthetic *a posteriori*.

Philosophers do not agree on the use of *a priori* and *a posteriori*, and this treatise will not attempt to referee disagreements. 'Kant claimed that synthetic *a priori*, truths were to be found in two fields-mathematics and the presuppositions of experience and science. He maintained that although mathematics did contain some analytic truths (since these were propositions which summed up purely deductive steps), the main bulk of mathematical truths were synthetic *a priori*.' (Edwards. 1972. 1:140) Analytic statements cannot be false. (Edwards. 1972) To add to the confusion, Quine, W. V. (1908-) who studied under Whitehead, maintains that the analytic-synthetic distinction has not been made. (Edwards. 1972. 7:53) [Bunge, M. 196l. in Mind, defends the notion of analytic *a posteriori*, one of few to do so] [Quinton, A. 1963-4. in PAS, along with several authors, rejects synthetic *a priori* .]

ANALYSIS-SYNTHESIS

Analysis is very important and fundamental to synthesis. Analysis has provided information, data, and critical thinking without which there wouldn't be 'pieces' for synthesis. Analysis will continue to be important but cannot be depended on to continue to make the same impact or contributions in the future as in the past. Synthesis will play a major role in the future, often serving to guide investigators and thinkers toward the most important or needed analysis. So extensive has been the use of analysis that decision-makers often decry, perhaps unfairly, paralysis by analysis. Analysis has distinguished Western society from Eastern society in many ways.

Although analysis and synthesis are often considered as opposites, they should, in fact, be considered as complementary. Information flows back and forth between analysis and synthesis. The brain is like a membrane between the two sorting from infinite sources of information to enhance a unity of purpose. (Image 3-7)

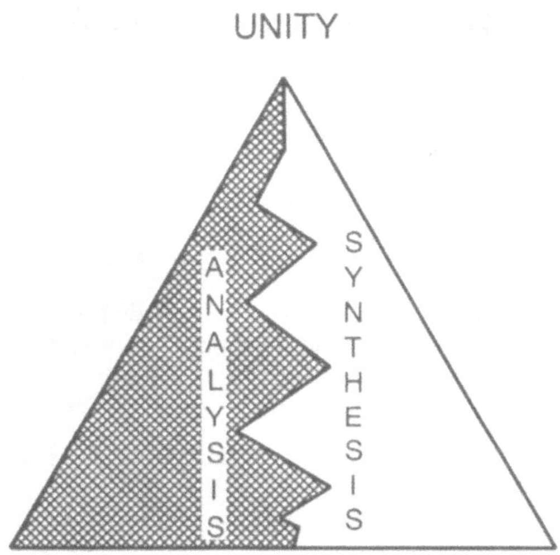

Image 3-7. Melding of Analysis and Synthesis into Unity

Thinking may be of the analysis-synthesis or synthesis-analysis approach. Analysis is a method that can be duplicated by others, step-by-step, to determine whether the results are verifiable. Analysis provides an approach to evaluate synthesis as a first step basis, but certainly does not provide an in-depth evaluation based on the objectives of synthesis. An important aspect of evaluating an artifact of synthesis is the response of the user—who must either be convinced or become convinced of the merit of the artifact. The work of an artist whether a painting, poem, or literature must satisfy a user—although it may require waiting for another generation for favorable judgment. Thus, a building, automobile, or device, in addition to withstanding the scrutiny of critical analysis, must also meet the approval of the consumer or user. The uncertainty of the response of the user or consumer places a product of synthesis in a different category than the more predictable or repeatable result of analysis.

SYNTHESIS IN OTHER CULTURES

Synthesis is represented as a more holistic thinking in Eastern cultures, where unity is sought, and in the Native American culture, where harmony with nature predominates. Synthesis is not new. Finding substitute materials, such as ceramic and glass for metals, or devising new systems, such as improved bows and arrows, and considering a larger 'whole', such as the environment, are longtime approaches. Knowing the shortcomings of their system, the Japanese developed Departments of Precision Engineering in the late thirties with German assistance. The shortcomings of their system to provide for precision is an indication of the weakness of analysis in their thinking. The major contribution after World War II by Dr. W. Edwards Deming, after whom the world-known Deming Prize is named and funded by the government of Japan, was for strengthening the role of analysis in their thinking in a very productive, effective manner. Dr. Deming has since developed fourteen principles of management that integrate analysis and synthesis.

An example of Japanese thinking in a synthesis mode is illustrated by the title selected by Toyota Motor Corporation for the quarterly magazine published for general distribution. The title 'the wheel extended' leads one to think in terms of a larger system (without specifically saying so) of which the wheel is a part, not only a part of the automobile but of society as well. The wheel is the Buddhist symbol of unity and wholeness. In the U. S. when the American Society for Engineering Education (ASEE) developed a new publication for engineers, instead of using a title that represented the synthesis nature of engineering, used the title 'prism' which is generally a tool of analysis. The prism can be used to break the light spectrum into its components, although the principles involved in the understanding of light were originally elucidated based on synthesis, which could be the meaning of the title.

CAUSES AND EFFECTS

The words 'causes and effects' do not appear in science journals with regularity, particularly in the physical sciences. (Image 3-8) In journals in which the sciences are applied to practical purposes, and in works of engineering, there is considerable references to 'cause and effect.' In medical journals the use of 'cause and effect' is common.

If something doesn't work properly (the effect), or for one who is sick, we ask about the cause(s). The search for the cause(s) is a diagnostic procedure, whether for a natural or unnatural system. Causes are among the antecedents of the effects. The word 'cause' is at home in the diagnostic and of the so-called applied sciences and the professions, rather than in the physical sciences. There is no need for the term 'cause' to explain within theories themselves. Toulmin (1960) uses the coparison of a map and an itinerary to illustrate the comparison; travelers in former years relied more on an itinerary than a map.

Aquinas explains from causes to effects as *a priori* and effects to causes as *a posteriori*. Descartes describes analytic as effects from causes and synthetic as effects to causes. Newton describes deduction as causes to effects and induction as effects to causes.

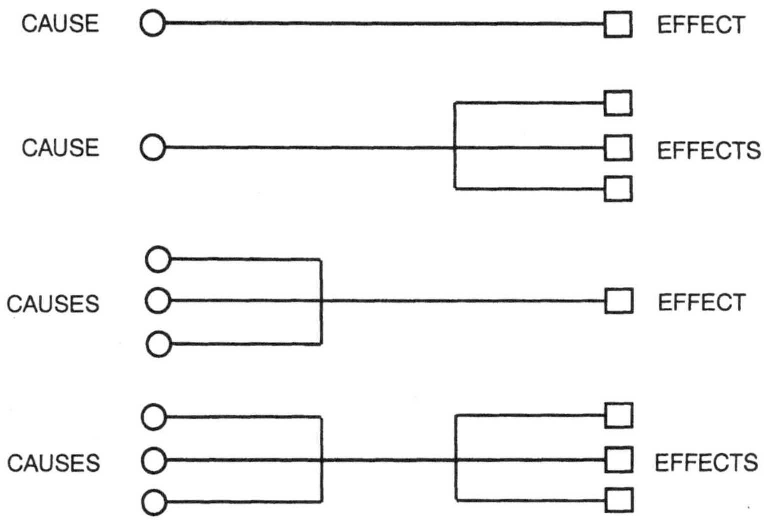

Image 3-8. Causes and Effects

The approach to determining causes and effects are much the same for natural and artificial systems. The knowledge base for natural systems, using laws of nature, may be helpful for both systems. The knowledge base for artificial systems may not contribute much for understanding natural systems except as models of behavior and response.

In a biological system, a human being for example, the interrelationships between and among systems (perhaps more appropriately identified as subsystems) are very complicated and difficult to isolate to determine the causes and effects. For example, there are many possible causes of the effect blurred vision. Astigmatism, cataracts, detached retina, eyestrain, glaucoma, tumor, chemical unbalance such as with diabetes can be considered as a possible cause. There could be a less obvious cause, or a combination of causes involved, making diagnosis difficult.

In an unnatural or artificial system, such as for an electric power system, one can consider (diagnostic) the various potential causes for a power outage. The process would normally be straight-forward if there is only one cause. When two or more causes occur simultaneously, the identification of the causes is usually quite difficult to discern. A blown fuse, broken power service line, failure in a transformer, damaged distribution and transmission line, failure of the generator, malfunctioning of a controller, or the source of energy driving the generator, each can be considered as a cause. There could be a less obvious cause, or a combination of causes, making diagnosis more difficult.

David Hume (1711-1776) offers three descriptions of cause: (1) an object precedent and contiguous to another, and where all objects resembling the former are placed in like relations that resemble the latter, (2) an object precedent and contiguous to another and so united with it that the idea of the one determines the mind to form the idea of the other, and (3) or in other words where, if the first object had not been, the second had never existed. (Edwards, 1972)

SCIENCE AND REASONING

The science community has been built on deductive and inductive reasoning, and to some extent intuitive reasoning. Deductive-inductive thinking applies to the sciences, but generally not to the arts and professions. Analytic-synthetic thinking applies to the sciences; synthetic-analytic thinking applies to the arts and applied fields. A change to analytic-synthetic or synthetic-analytic thinking with their common linkages would accommodate the arts, sciences, and professions, and help reduce the "two cultures" syndrome, expressed so eloquently by C. P. Snow (1962). He writes of the lack of common ground between the two groups—the literary intellectuals and the scientists, particularly the physical scientists. There is a lack of understanding and mutual incomprehension, and even a difference in attitudes. He claims that the creativity of both groups, sciences and the arts, suffer as a result of the separation of the two cultures.

Combining analysis and synthesis (a kind of anal/syn-thesis), initiated as a way of thinking, as well as a way of doing, offers one constructive way of linking the two cultures of arts and sciences into one, and accommodates the learned professions and applied subjects as well. Future needs demand that more attention be given to teachings related to synthesis in our educational activities.

It is generally agreed that many young people lose their previously exhibited creativity and perhaps the capability of synthesis as they move through the educational process in the United States. Educators need to incorporate more activities which recognize synthesis as an important characteristic needed for world citizens of tomorrow. This is not necessarily a new or additional course, but a way of thinking, a course of action, and a perspective in current and new courses and experiences of students. These approaches apply not only to those who enter advanced learning for the arts, sciences, and professions, but to those who vote and live in a democracy. As government gets involved in numerous ways in controlling the artifacts (products) of synthesis, as well as the process of synthesis, an informed citizenry is essential, so that thinking is more responsible. Although not proven, there is the potential that a person strong in synthesis would have views that would counter the selfish views of individuals as described by Ayn Rand. A person with views of a synthesist would be expected to consider a larger system and take a longer view than commonly done by an individual looking inward only.

SUMMARY

Thinking is the beginning of intellectual activity. Methods of thinking, reasoning, and dialectical argument are analytic and synthetic. Intuition plays a major role in the advancement of thinking, but is unpredictable and probably can't be taught.

Paradigm changes are more likely to result from breakthroughs in synthesis as compared to analysis. One immersed in analysis who has a broad view and an open mind and searching thought process will provide breakthroughs that could be classified in the category of synthesis. Business, educational, and government organizations often have analysts as employees. In the future, synthesists will occupy positions of importance in organizations to provide their special expertise.

Different philosophers present different views on *a priori* and *a posteriori* statements, but there is general agreement that the first phrase as being analytic and the second statement as being synthetic.

A combined approach of analysis and synthesis with continual interaction of the two, provides an approach that could minimize the 'two culture' syndrome of the arts and sciences described by C. P. Snow. Identification could highlight the combined approach, such as anal/syn-thesis, and lead to more rapid integration not only of arts and sciences, but of the professions and applied fields as well.

Synthesis needs to be considered as a way of enlarging one's knowledge and concerns to be a more informed citizen. One with a greater sensitivity and capability for synthesis could have a greater concern for the welfare of others in the system under consideration. Synthesis would help individuals put more emphasis on developing character in contrast to strengthening personality, both of which are important aspects of a person. Recent generations of students and the adults resulting from those students appear to have put greater emphasis on developing personality which could short-change people from their potential and limit the development of society.

Part 4. Synthesis in Context of Current Developments

PHILOSOPHY APPLIED

Early philosophy pertained to thinking about religion on questions that could not be quantified. Philosophical judgments are synthetic rather than analytic and mathematical thought is analytical. Proof consisted of convincing others by persuasion. Thought, logic, and words were the weapons of persuasion. Some will argue that as soon as an issue or question can be answered by data, facts, and information that philosophy is no longer involved. In that sense, there isn't an applied (or practical) philosophy. But, as in other fields, a new theory will stimulate others to try to prove or disprove it, as Einstein's theory of relativity stimulated at least a generation of scientists to prove or disprove his theory. Einstein expresses the important role of synthesis when he wrote, "It is theory which decides what we can observe." Leonardo da Vinci's sketches, the synthesis of ideas for artifacts, such as the parachute and the helicopter, are still being studied by analysts, even today using the supercomputer. The philosophers deal with thoughts. They deal with thoughts and ideas beyond the physical. Philosophy could be considered between religion and science in terms of realism-religion is based on faith and science is based on proof of existence. The scientists have normally dealt with things as they appear in nature, but more recently have moved toward dealing with things resulting from their work. The engineers, architects, and artists deal with human-constructed or made-things, manufactured items, and the unnatural or artificial as contrasted to the natural.

So the thoughts of the philosopher become expressed as a thesis (speaking and writing) that in turn become things, either in natural systems for the scientist or in built-, constructed-, or artificial-systems for the engineer. The20th Century witnessed major advances in the physical world with an emphasis on analysis. Analysis gained in momentum through the 19th and 20th Centuries. The evidence is mounting that the 21st Century will be characterized by synthesis, not only in thought (thinking) which has been going on for several centuries, but also for the artifacts (things) of human beings.

INCREASING EMPHASIS ON SYNTHESIS

The evidence is that there is an increasing emphasis on synthesis. Writing in science and engineering and literature use the word synthesis and related words to illustrate a greater connectivity of subjects and concerns. Commercial concerns advertise their capabilities and strengths in terms of synthesis, often using syn-... to emphasize that capability, as for example, synectics and synoil.

Synthesis, as expressed throughout, is an old concept. However, we now have an array of new tools, based on computers and the software to use them. The computer is an important tool to help think more deeply through analytic extension, but through multimedia software is an important tool to consider the broader approaches associated with synthesis. Large databases on different subjects, high speed interactive databases, decoupling of data from former roots or bases, and creative modeling and programming make synthesis a realistic approach previously impossible. The global communications network provides the vehicle for involving knowledge and experience, archival and present, to support the synthesis approach.

Larger systems are being built in practically all endeavors of our economy from banking to communication to transportation requiring more comprehensive thinking and a synthesis of information from several fields. The constraints of the human (often overpopulated, lacking in needed skills, improper nutrition) and natural resources (minerals, intensity of resource use, environmental impact) in which much of the modern world is focusing on Western progress, the connectivity of systems must be considered and appreciated.. The natural environment in relation to the 'engineered' environment and its relation with energy, with materials, with economic resources, and with ethical concerns, require a more holistic approach. The integration of ideas, organizations, means of communication and information to improve the system(s) is receiving considerable attention by various groups in society. Integration is a step toward synthesis. There is an increased desire to meet user demands (wants and needs), which involves synthesizing systems to optimally and economically meet those demands. Examples are just-in-time (JIT) manufacturing, agile manufacturing, total quality management (TQM), computer integrated manufacturing (CIM), and flexible manufacturing systems (FMS), customer oriented management systems (COMS), and management information systems (MIS).

There are increasingly more options for synthesis, often the result of increased results from analysis. Instruments and computers add to the capability of those interested in synthesizing engineered biological products, in considering the human systems relating to the physical systems, including education, health and nutrition, economic, political, and educational systems.

MEANING OF SYNTHESIS AND EXAMPLES

The general meaning of synthesis based on the etymology of the word is to put together to form a whole. In philosophy synthesis is reasoning from the particular to the general. Synthetic items are represented by words such as artificial, substitute, composite, imitation, and simulated, as well as synthetic. There is no systematic, standardized, accepted use of these terms. At some point of complexity, yet ill-defined, an item is called a system. Although a composite material is a system of elements and compounds, the term system is usually reserved for items with separate acting but interrelated parts.

Humans have sought new and improved physical tools since the beginning of history, representing new systems with the development of new tools. Humans are often distinguished from animals by their ability to conceive and build tools to meet their needs and wants. Animals such as birds build their nests based on innate capabilities, but the nests do not evolve into a new form, although different materials might be used depending on availability, the design remains the same. Beavers build dams in much the same way that birds build nests. Humans can learn from these animal actions, and only slightly do animals learn from humans, except for domesticated animals.

EARLY SYNTHESIS

Synthesis of natural materials is a practice less than 200 yr. old. Urea $(CONH_2)_2$ CAS Registry No. [57-13-6] was the first natural organic compound made by man by Friedrich Wahler (1800-1882) in 1828. This was followed by the development of synthetic dyes by W. H. Perkins (1838-1907). Synthetic urea, manufactured by combining ammonia and carbon dioxide under appropriate temperature and pressure, is widely used in commerce and is a feedstock for animal feed, fertilizer, and some plastics.

Substitutes (artificial) for biological materials, particularly for food and pharmaceutical purposes are of more recent vintage as illustrated by the invention of margarine.

> "In 1939 we celebrated the 70th anniversary of the invention of margarine. It was 1869 that Napoleon III commissioned the chemist Mege-Mouries to produce a butter substitute that was to be cheaper than natural butter. Since then, technical progress has smuggled into our fare an endless number of substitutes, synthetic concoctions, and artificial products."
>
> *(Juenger, 1949)*

USE OF THE WORD ARTIFICIAL

The words artificial and synthetic are often thought of as interchangeable. Use of the word artificial as something synthetic may be true in some cases, but its extensive usage in that manner will probably continue, even though misleading and inaccurate. R. B. Fuller is rather insistent of his view on the subject:

> "A plastic fender in place of a steel fender is no less a fender because it is made of plastic. Its properties are different but it's still a fender."
>
> *(Fuller, 1979)*

Also, the fender would not be considered as a system, often considered as a result of synthesis or the basis of analysis even though it is made up of several components, but those are not interrelated components. Perhaps it is a materials system. Just as a basic element, such as iron in chemistry, is not a system, a basic part is not a system.

> "In my viewpoint, there is no meaning to the word artificial. People can only do what nature permits them to do ...People make discoveries of principles operative in nature and often find ways of generalizing those principles and reapplying them in surprising directions. That is called invention. But they do not do anything artificial. Nature has to permit it, and if nature permits it, it is natural. There is naught which is unnatural."
>
> *R. B. Fuller. (Wagschal and Kahn, 1963)*

Or take the case of making snow by forcing water under pressure through nozzles in a cold atmosphere to form snow. It is called artificial snow, yet the snowflakes look like the snowflakes made naturally. It is certainly not synthetic snow. It is called artificial snow, but there is nothing artificial about it. The snow was mechanically-made, perhaps artificially-made, but not artificial. The term unnatural is more appropriate designation.

REPRESENTATION OF SYNTHESIS

Synthesis is represented by numerous examples as presented in Matrix 4-1. Generally, some action is required to cause a product to be formed. These actions include:

» To combine varied and diverse ideas, materials, components into a coherent, consistent, often complex, product or system, to produce an artifact.

» To assemble elements to make a compound or mixture already known based on a previous analysis or knowledge. Some elements may be omitted, or substituted by other elements or by substituting different elements, making a new substance.

» To assemble elements or substances not necessarily in the original substance to provide an end product with end use characteristics similar to the original substance or enhanced from the original substance. Synthetic fibers could be man-made textile fibers made from natural materials such s rayon or acetate. The use of nylon for silk, that is substituting nylon fibers for silk, is considered as synthesis. Synthetic rubber made from neoprene or butyl resembles natural rubber. The products are sometimes called simulated products.

» To make new products that substitute for other products without any of the original elements. Thus, artificial sweeteners are not primarily food sugars but other components, such as calcium saccharin [2634-33-5] and aspartame [73537-92-5], substituted to give an effect similar to sugar.

The same is true for table salt [7647-14-5] substitutes, consisting of other chemical components, such as potassium chloride [7447-40-7], monopotassium glutamate, and glutamic acid [56-86-0]. The use of margarine for butter as a substitute spread could hardly be called a synthetic product, but margarine is a vegetable oil that has been processed to provide the characteristics of butter. A hot dog made out of turkey has substitute food materials but probably shouldn't be considered synthetic, although new components have been assembled. A toy train made out of wood or plastic is a model of a train, thus simulates a train.

» To utilize a natural biological system to provide growth, reproduction, or continued life through photosynthesis

» To produce by a manufacturing process, such as materials processing, new materials, new products for materials with specific properties to meet certain performance requirements. These new manufactured materials are often called composites, unlike the materials available naturally.

A special concern is expressed for synthetic minerals, vitamins, and foodstuffs. Synthetic elements and compounds, such as calcium for food and folic acid [59-30-3] for vitamin B, accurately include the principal component of the natural substance, but may not provide the same effect in the animal and human body. Calcium, whether from a mineral or vegetable source, is calcium, Ca. The naturally supplied calcium in a food may carry along with it other important elements, such as magnesium and phosphorous, or compounds, for a favorable response in the body. The naturally supplied folic acid is a major component of Vitamin B, and both might be required for a favorable nutritious or health result. The synthetic product would normally contain the same and the majority of components found in the natural product it represents, but be lacking or short of critical minor elements. Minor components may have a very important role in utilizing the major components, being a catalyst or providing appropriate chemical linkages. This is not to say that the synthetically produced material is inferior to the naturally produced. In some applications the synthetically-produced might be superior, particularly where a pure element or compound is required. The chemical formula of the compound is not sufficient to say that a particular substance is the synthesis of another, such as lactic acid, $C_3H_6O_3$ [598-82-3]. As an example artificially produced lactic acid will turn polarized light to the left, and the naturally produced lactic acid will turn polarized light to the right. The first, levolactic acid, can be toxic and the second, dextrolactic acid, is a food (sugar), yet both have the same chemical formula. The interrelationship between minerals and vitamins suggests the need for a view of the nutritive system above (broader) that of the food elements by themselves, so that a higher level (hierarchal) of synthesis is involved to obtain and to meet a need beyond the basic elements themselves. Water, chemi-

Matrix 4-1. Examples of Artificial, Imitation, Simulated, Substitute and Synthetic as Used

Item	A	B	C	Representation in practice
Artificial food color			•	using natural food materials to color
Artificial heart			•	biocompatible materials
Artificial insemination	•			same process, but done artificially
Artificial intelligence			•	understanding of intelligent action
Artificial snow		•		same chemical content, made artificially
Artificial kidney			•	ultrafiltration membranes
Artificial sweetener			•	non-nutritive, saccharin, aspartame
Composite wood			•	impregnate wood with phenolic resins
Imitation gem		•		made to look like natural gem
Simulated		•		properties made to look like original
Substitute natural gas	•			methane made to replace natural gas
Synthetic chemical	•			use engineered bacteria-one approach
Synthetic crude oil	•			manufactured from coal
Synthetic detergent			•	replaced sodium in long-chain acids
Synthetic diamond	•			same chemical content as natural
Synthetic drugs		•		nearly like natural drugs
Synthetic dyes		•		plant material replaced by coal tar
Synthetic fuel		•		does not occur naturally
Synthetic gas		•		manuf. from shale or tar sands
Synthetic gem	•			same chemical, physical properties
Synthetic rubber		•		butyl and neoprene
Synthetic steroid	•			made by biosynthesis

A Same chemical makeup, product made artificially or by synthetic process

B Substitute product for a natural product (replacing most of the properties of product)

C New product with different chemical contents to perform function of original product

cally, is a combination of hydrogen and oxygen. Yet, water made by combining these two elements would not include many of the components normally in water, particularly drinking water, and would not be inviting to the consumer, nor as healthy in many cases.

THREE APPROACHES-SYNTHESIS

Synthesize/synthesis, the words, represent three approaches, with usage or practice determining how the words are used or applied:

Class I. Synthesize/synthesis is considering a perspective system, a larger whole, holistic view, as a basis of developing, inventing, understanding, to make, to build, to provide an entirely new "thing", such as a painting, statue, building, product, or system considering several alternatives for making that system. Design includes considering elements available or obtainable, putting those elements together, to be built, which until recently was a built- or constructed-device or system, in contrast to a natural system. In recent years with new knowledge in molecular biology, biological systems are being built, classified under the category of genetic engineering, an example of synthesis. Interfacing of natural and artificial systems provides another opportunity for expanding the use of synthesis in the future. A powerful use of synthesis in the future will be to expand the interrelation of physical and human systems.

Class II. Synthesize/synthesis is anticipating how known parts or components could be fit together in a coherent manner to meet an objective. The pieces of a jig saw puzzle can be assembled to make a picture, based on *a priori* knowledge, in which the assembler must fit the pattern of the designer; a toy set, such as Lego™ has numerous parts that can be assembled to make several different items, more to the imagination of the assembler.

Class III. Synthesize/synthesis is putting materials together in such a way to provide a new product or to replace a product, i. e., nylon for silk, soybeans for milk, margarine (vegetable oils) for butter (animal fat), saccharin for sugar, imitation coffee cream. Chemurgy, a word used in the 1930s, represents an item not of natural origin, artificial, devised, substituting and enhancing of a natural product produced by other than natural growth. Synthetic rubber may have the same chemical constituents as natural rubber, but made by manmade methods. Synthetic fabrics for cloth or carpeting may not have natural fibers involved and for specific uses may have properties superior than the original natural material. The representation becomes even more uncertain when one considers that humans may interfere in the breeding and selection of plants and animals to produce different results from the original natural item. Artificial insemination for breeding produces a regular offspring, so it is the process, not the product, that is artificial. A review article covering the published literature of several publications, after the fact (Class III), would have less influence over the subject than a planned review article based on participation of those authors before their publications are written (Class I).

The numerous words used in the literature and in spoken communication, provide a glimpse of the many applications of synthesis, as given in Matrix 4-2.

TWO APPROACHES-ANALYSIS

Analyze/analysis, the words, represent two approaches, with usage or practice determining how the words are used or applied.

Class I. Analyze/analysis which is done for the purpose of learning more about the parts and subparts. Although the use of the results is not limited, the use of the results for synthesis is not the objective.

Class II. Analyze/analysis which is done for the purpose of using the results for a larger system, which contributes to synthesis and can be used both for contributing to a synthesis/synthetic system or evaluating a perceived or real system.

IMPACT OF ANALYSIS

As a result of the intense use of analysis an expression has developed, "paralysis by analysis," an indication that analysis is being carried out for the sake of analysis beyond that needed for productive results. By imposing synthesis on the process some limitations and boundaries could be established to help guide the analysis. The detail or precision of data being obtained by an analyst, with additional cost for greater precision, may be unjustified when one considers the overall precision of the larger system into which the data are to be used. The larger system being synthesized could be used as a guide in determining the limit of useful precision for the particular system under consideration.

Analysis can often be accomplished by individuals whereas synthesis, particularly of systems, usually involves several people, often with an organized group or panel. Often one person through synthesis is the source of a creative approach or idea—that's their way of thinking. Creativity can be encouraged and supported in both analysis and synthesis. Although the group may tend to limit the creativity of some people, the members of the group can also be the source of ideas or stimulate new or additional ideas.

Analysis has a very important role as a part of synthesis. Analysis and synthesis have complementary roles and are not usually separate. Without analysis, synthesis would not be as useful or effective. With synthesis analysis can be even more effective. Analysis can be one method for evaluating synthesis, but is not sufficient.

Synthesis is an effective process to utilize for responding to what can be

Matrix 4-2. Synthesis and Near Synthesis Synonyms

a
aggregate
amalgamation
arrangement
artifact
artificial
assemble
assembler

b
Baconian
biosynthesis
blend

c
chemosynthesis
chemosynthetic
coalesce
coalescence
coherence
combination
combine
compare
complete, completeness
compose, composed
conceive
conception of
  conjunction (Kant)
concommitant
connectivity
connexity (Whitehead)
copy
cross-disciplinary

d
design
designed

e,f,g,h
electrosynthetic
engineer, engineered
episynthetic
fabricate
fabrication
fashion
forge
formulation
fusion
generalization
harmonize
harmony
holistic
horizontal thinking
hypothesis

i,j,k,l
image
imagine
imitation
imitative
incorporation
indiscernibility
indiscerptibility
induction
inference
integrate
integration
interconnectivity
interdependence
interdisciplinary
intersubjectivity
intuition
invented, invention

l,m,n,p,q
manmade
megasynthetic
merger
mixture
monosynthetic
multidisciplinary
omneity
omni-integrated
  synergy (Fuller)
oneness

osteosynthesis
ought-to-be
pan-disciplinary
parasynthetic(al)
photosynthesis
polysynthetic(al)
produce
production
psychosynthetic(al)
put-together

r,s,t
reproduce
speculative
symbiosis
syncreation
syncretism
synergy
system
systemization
synthesis
synthetic
trans-disciplinary

u,v,w,x,y,z
unification
unify
unifying
union
unity
universal
whole
whole emnity

imagined. Synthesis provides a path to pursue the imagined idea or artifact to a greater extent than starting with analysis, which may inhibit the possibilities of favorable results before it is given a chance to materialize. Examples of the wide use of synthesis in the literature of arts, professions, science, and technology are listed in Matrix 4-3. Companies are using a syn- word in their titles or in the description of their activities too indicate the emphasis of synthesis in their work.

ARTIFICIAL SYSTEMS

If taken to the smallest element, all items except atoms themselves could be considered as systems, and even atoms could be divided into components. Thus, atoms are parts of compounds, compounds make up chemicals, and chemicals make up materials systems. We consider an apple as a food item, but we refer to several foods as fruits, vegetables, meats, beverages, and desserts as making up a food system. Although several elements or parts make up an automobile, we don't refer to it as a system which would be perfectly true, but it is a part of the transportation system. An artifact (artificial) is a made item, such as a painting, skyscraper or bridge. The painting could be called a color system and the bridge a structural system and a part of the transportation system. In the natural domain a flower is certainly a system of components, but in practice we wouldn't refer to a flower system, probably because the interconnectiveness is not physical, but would refer to a landscape with a system of plants. To avoid diverting the more central issue of synthesis no attempt is made to define separately those items, natural or artificial, that are commonly called systems and those that are not.

Matrix 4-3. Syn- in the Literature, Representative of Several Fields (not including numerous chemicals beginning with syn-)

Word	Use	Source/reference
Synapse	Neural junction	McG-H
Synchronization	Motors, clocks, communication	Meyers. 1987; McG-H
Synchros	Robotic sensors	Meyers. 1987
Synchrocyclotron	Frequency modulated cyclotron	E. B.
Synchroflash	Photography	E. B.
Synchromism	Modern art movement	E. B.
Synchroneity	General	E. B.
Synchrotron	Communication, noise, astronomy	Meyers. 1987
Syncopation	Music	Webster 3rd Int. Dict.
Syncretism	General	Webster 3rd Int. Dict.
Syncrude	Petroleum product	E. B.; K-O; McG-H
Synderesis	History	Greene. 1991
Syndiotactic	Materials, polymers	K-O; McG-H; Meyers.1987
Synecology	Environmental relationships	Herit. Dict.
Synepistemic	Transdisciplinary	Klein. 1990
Syneresis	Synthetic membranes, colloids	E. B.; K-O; McG-H
Syngamy	Nuclei of sex cells fusion	Wilson. 1975
Synizesis	Chemical gel	Herit. Dict.
Synoptic	Meteorology	McG-H
Syntactic	Computer programming	Myers. 1987
Syntagma	Elaborate philosophy of Eucken, R.	Kiernan. 1965
Syntax	Computer programming	Singh. 1987
Synthetase	Bacteria	McG-H
Synthol	Fuel conversion reactors	K-O
Syntype	Taxonomy	McG-H

Sources: (E.B.), Encyclopaedia Britannica. 1987. 1993; (K-O), Kirk-Othmer. 1984; (McG-H), McGraw-Hill. 1987; Rigney and Barnes. 1980; plus those listed above.

BRAIN AS A SYNTHESIZER

The brain is a powerful synthesizer and integrator of knowledge and information. A popular article in the New York Times (Tues. Sept. 10, 1991, p. C1), reported on a theory developed regarding the functioning of the brain. It is believed that the brain [does not have a filing cabinet for storing concepts, but the brain instead] collects information from various storage sites and transmits the information to a separate center to provide information for the concept.

When a person thinks of a cup, clusters of neurons that store knowledge about a cup (container, material, color, liquid, texture, breakable, etc.) project the information into a common convergence zone.' The concept develops as a word is said, not stored ahead of time. It appears that the brain responds to purposeful intent, much like the process of synthesis in the physical, particularly unnatural domain. Experiments reveal an ever increasing finer subdivision in the brain. How the brain takes bits of information on various subjects to create a synthesis and unity of thought is known
as the binding problem. These concepts learned may be the basis of developing a stronger appreciation and use of the brain for synthesis. We know from experience that the brain is highly developed to handle analysis (left brain) and those abilities can be more strongly developed.

Also, interestingly, it is believed that different parts of the brain are used for natural and manmade objects. There is a simultaneous firing of widely distributed neurons—it is believed that the system mechanism provides feedback firing from the 'convergence zone' and that the brain creates unified circuits by oscillating distant components at a shared frequency.

The fact that the brain responds differently to natural and manmade objects or systems may lend credence to the thought that different consideration must be given to natural systems (science) and artificial systems (engineering and others). At the same time, this theory of operation of the brain puts considerable emphasis on its function in synthesizing, that is, pulling information from various sources to fulfill the Image or concept. This implies a difference in THOUGHTS and THINGS as mentioned previously.

The brain is the basis of the remembered, that is for habit and innate activities as well as thinking and thought. The brain of humans provides a capacity beyond that of animals, so that the larger brain is available 'for imagination, abstract reasoning, and creative fancy, in addition to the use of the brain for controlling bodily functions, plus the fact that humans have much more surface area of the brain because of its ridges and valleys.' (Asimov, 1987) Like the limitations on describing synthetic food components, it is undoubtedly true that more than size, surface area, and ridges and valleys of the brain are required to represent the differences between human and animal brains.

LEFT BRAIN AND RIGHT BRAIN

It is generally agreed that the left side (hemisphere) of the brain is specialized for analytic functions, language being only one of the functions, and that these functions are the result of the superior analytic skill. Likewise, the right side (hemisphere) of the brain has superior spatial capabilities that are derived from its synthetic holistic way of dealing with information. (Springer and Deutsch, 1985) Studies of the neural system are becoming useful in the design and development of computers and control systems, using the functioning of the natural as the basis for advanced unnatural or constructed electronic and mechanical systems. The literature of neural nets (networks) is increasing rapidly. In 1987 there were fifteen references on the subject in the Applied Science and Technology Index; in 1992, there were 175 references.

FUZZY SYSTEMS

In 1965 fuzzy sets and fuzzy systems were developed as a means of describing and representing complex relationships of real world phenomena with considerable uncertainty in mathematical terms. The approach provided a way of representing complicated relationships that mimic the input-output of the system without knowing or without using known detailed relationships of the components of the system. Fuzzy logic utilizes the computer to obtain data from sensors, to evaluate data and information and to respond. Fuzzy logic 'learns about known objects and processes in the supervised learning mode, then stores and grades the characteristics for use later in the unsupervised projected mode. The procedure provides a means of handling considerable data, often not precise but representative, to utilize that data to provide a response such as a command and control mechanism. Thus, an irrigation system can be directed to operate at the most economical time in response to weather and soil conditions and changes, as related to crop growth. Fuzzy logic can be used to provide instructions to cars so that they can park themselves. Fuzzy logic can

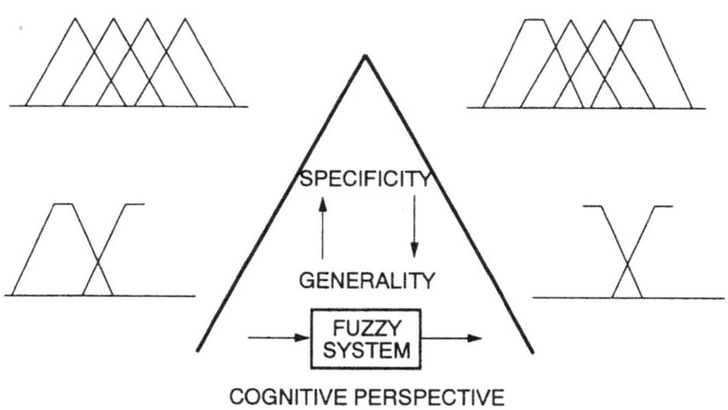

Image 4-1. Visualization of a Fuzzy System (Pedryez, 1990)

be used to control manufacturing operations to minimize waste, optimize profit, optimize use of natural resources, reduce pollution, or other functions it is possible to optimize.

That approach is characteristic of neural-engineering and exemplified by the Hopfield Model. The model does not duplicate neurons, necessarily, but provides an approach to represent what neurons do. The technique developed represents a synthesis approach to developing the models of systems, and an analysis approach to test the model. Pattern recognition, control relationships, knowledge engineering, and expert systems are useful subject areas for applying the principles of fuzzy systems. Thus, the strict application of analytical mathematical tools, which is very difficult for complex phenomena, is replaced by a synthetical approach. The synthetic approach could lead to improved analytical tools, if needed or warranted.

LUMPED PARAMETERS

A substance, material, or variable might be represented by numerous parameters, These parameters could be numbers that represent a particular characteristic. For many applications, although it would be desirable to have the parameter of each variable, in many cases those parameters are not available. Often these parameters can be combined in a generalized form, lumped into one parameter, to adequately represent the relationships at hand. The generalized form simplifies the equations representing the material or relationships, making an acceptable result or solution. Judgment is needed as one synthesizes the approach. The results must be used to evaluate or validate the applicability of the lumped parameter, one number representing several parameters.

POST-INDUSTRIAL SOCIETY

Bell (1973) foresees the general arrangement of social and technological change from a pre-industrial to industrial, to post-industrial society. (Matrix 4-4) These

Matrix 4-4. Schemes of Changes from Pre- to Post-Industrial Societies

	Pre-Industrial	**Industrial**	**Post-Industrial**
Time perspective	Orientation to past	Ad hoc adaptations	Orientation to future
Economics	Extractive agriculture mining fishing timber	Goods manufacturing building processing assembling	Transportation, utilities, trade, health, recreation, services
Technology	Raw materials	Engineering	Information
Methodology	Common sense Experience	Empiricism Experimentation	Theory, models, simulation, systems anal.

Adapted from Bell, Daniel. 1973. The Coming of Post-Industrial Society. p. 117.

projected changes parallel those changes enhanced by thinking synthetically, and producing synthetic (artificial) things and systems. And like synthesis that builds on analysis, so the post-industrial society builds on the pre-industrial and industrial society.

SUMMARY

From early history, philosophy was the major 'thinking' activity, primarily as related to religion and government. During that period 'thinking' was also devoted to natural systems. Later, 'thinking' was directed to unnatural or artificial systems. The earliest of the 'thinking' in philosophy was by synthesis thoughts, except for a few leaders physical systems were studied analytically, and now we are entering a period in which 'thinking' synthetically is gaining importance as a result of many unnatural things. Unnatural things, often patterned after natural things, or the result of an imagined thing, are the basis of many current developments. Both in the thought process and in the product provided, synthesis is at work, and the product is often called a synthetic product. Unnatural things as represented by artifacts; artificial, imitations, substitutes, and synthetics are in themselves the subject of considerable study. Unfortunately, there is no common or systematic usage of these terms and some terms are used interchangeably, leading to public confusion.

Using synthesis as an approach stimulates creativity, expands the horizon, incorporates more systems, such as improving human welfare that includes safety, environment, and health. The way the brain is believed to operate gives an example of synthesis of thoughts, emphasizing integration. To handle large amounts of data, new mathematical tools have been developed, such as fuzzy sets and lumped parameters, in an attempt to treat the whole process, rather than parts or samples of the process. Terminology relating to synthesis has become a part of management approaches and advertising schemes.

Synthetic products, particularly in the biological area, may include the major constituents of the original material, but be short of some minor chemical. The response of the synthetic product may differ from the original, particularly in a biological system, because of some minor difference, originally undetected. An antiknock component in gasoline fuel, a very small amount, like a minor element in a food or medical substance, can make a tremendous difference in its effect.

Note: Numbers in brackets [ ] are CAS Registry Numbers. CAS is the Chemical Abstract Services. The number identifies a specific chemical and is a reference to the chemical in published literature. More than one number may exist for a substance. Thus, there are situations in which a number cannot be assigned unequivocally to a particular substance.

Part 5.
Trends in Synthesis in the Arts and Sciences

The arts and sciences includes the subunits that define disciplines, such as art, botany, chemistry, literature, mathematics, and music. These subunits come together to represent the philosophics, the theory and principles, on the one hand and the technologics, the reduction to practice, on the other hand. (Image 5-1)

Recognition of the need to combine the efforts of disciplines has led to formation of interdisciplinary, multidisciplinary, and cross-disciplinary teams and groups to work on questions and problems in which the subject matter, available expertise, or user needs require such an approach. In addition, consortia of institutions are often formed to work on those subjects that cut across many functional lines. (Image 5-2) (Matrix 5-1) Synthesis is often involved in considering projects, problems, and questions involving more than one discipline and certainly for consortium-identified subjects.

ARTS

The arts and sciences greatly influence the intellectual development and the range of activities of the professions and applied fields. The arts and sciences also draw upon those professions and applied fields for methods, materials, and instruments to a much greater extent than usually acknowledged. Many of the advances in astronomy, materials, and chemistry could not have occurred without new instrumentation, and more recently, without computers. In art, new materials and paint media applicators, have increased opportunities for creativity. The synergy between fields increases not only as the options are increased by analysis, but as people recognize that these additional options increase the opportunities for synthesis.

The sculptures of primitive peoples, until recently considered primarily of ethnical interest, are seen as works of art. The sculptures represent what is seen or what is imagined, often a kind of synthesis in thinking that incorporates religion, life as it existed, family, and historical events. A sculpture is not necessarily a representation of a person or thing, but the artist's interpretation, which can be quite different from the view of another person, an approach imbedded in the ancient works. These primitive artisans utilized a synthesis approach that may have helped in establishing the Modern era.

In general, particularly in the last 100 years, Modern eras are dominated by analysis, and Post-Modern eras are characterized by a marked increase in synthesis. This section focuses on the role of synthesis and emphasizes synthe-

Matrix 5-1. Combined Disciplinary Activities

X-disciplinary	Head	Unit	Usual number	Unique features
Subdisciplinary	Principal Investigator (PI)	Section Sub-Unit Innerdepart.	1	Specialty within a discipline
Disciplinary	PI, Leader	Department	1-2	Discipline dominates
Interdisciplinary (interactions of	Co-PI, PI, Coordinator Leader	Interdepart. Program Team	2-3	Synthesis of two or more disciplines; Jointly publish in their disci-plines
Multidisciplinary (polydisciplinary)	Co-PI, Co-Director, Manager	Group Laboratory Organized Research Unit (ORU)	3 or more	Additive to discipline; Not integrative approach; Usually problem focused; In courses, students do most of synthesis
Cross-disciplinary	Director Manager	Center Institute	5 or more	Systems approach Applied, Partnership strategic interaction across disciplines; No intent of new discipline
Pan-disciplinary (polydisciplinary)	Director Manager	Consortium	Several	May be difficult to recognize discipline
Transdisciplinary (metadisciplinary)	Director Manager	Consortium	Unlimited	Overarching synthesis; More

Sources: Partially based on Klein (1990); Rubin (1986)

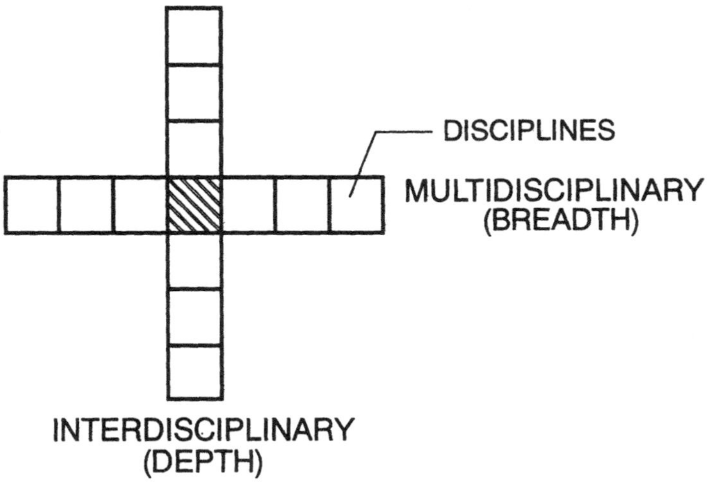

Image 5-1. Interdisciplinary and Multidisciplinary

sis-related approaches, to illustrate the increasing importance of synthesis while utilizing the strengths of analysis.

Art

The Modern Painting era began in the late 19th Century with the Impressionistic period giving way to the Postimpressionistic period. In the late 1880s, Paul Gauguin (1848-1903) and Emile Bernard (1868-1941) developed a philosophy and style of art that later was identified as synthesism, also called cloisonnism, that marked a change in approach from the previous Impressionistic period. The word synthesism was used by Gauguin to describe a style of art in which form is synthesized with the major thought or idea on the subject. The synthesism approach permitted an infinite variety of approaches in art. His forms became ideated, his colors suggested abstractions. His objective was to express an idea pictorially. A landscape seemed lacking in form as a result of being synthetically represented in blending colors.

The Postimpressionistic period in art, so named in 1911 but based on work done earlier, signalled a change in approach from analysis to synthesis. Previously the attempt was to reproduce the visual pattern on to the canvas. The first painters to be identified with this period were Paul Cezanne (1839-1906), Vincent Van Gogh (1853-1890) and Paul Gauguin (1848-1903). This new approach was a reaction to the strict boundaries of Impressionism in which the paintings were judged on the basis of their realism or likeness to the subject of the painting. Also, undoubtedly, the rapid development of photography forced

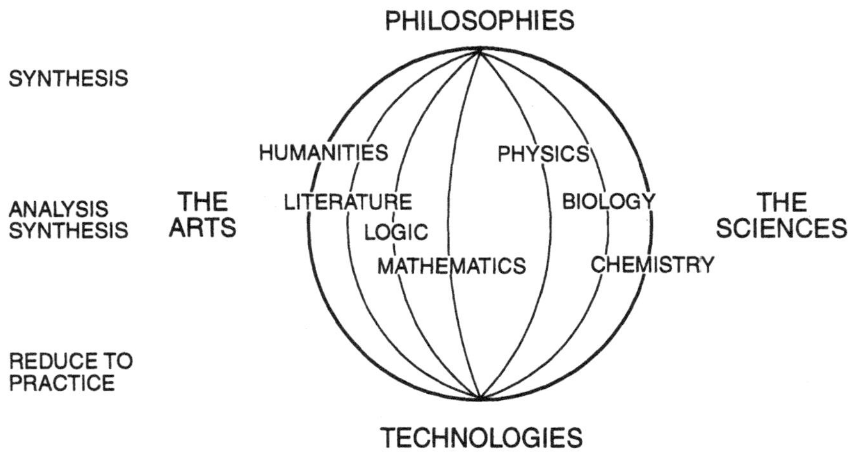

Image 5-2. The Arts-Sciences and Philosophies-Technologies Representation

a reassessment of the role of painting. During this new period artists took great liberties for incorporating imagination and using synthesis approaches, colors, shapes, materials, composition, visions, and images.

Cezanne's work became the basis of the Cubism movement, pioneered by Pablo Picasso (1881-1973) and Georges Braque (1882-1963). The Cubism movement, sometimes called Synthetic Cubism, was further classified into what is now called 1) analytical cubism (1910-1912) and developed into 2) synthetic cubism. In the latter, artists took greater liberties in use of colors, shapes, and textures. (World Book, 1990)

Pablo Picasso (1881-1973) is best known as a leader in the cubist movement He took conventional illustrations apart, then put the parts back together in unusual patterns, representing analysis first, then synthesis. Picasso continued to refine the way he represented and assembled the parts, the synthesis, which attracted the approval of many viewers. Through experimentation he found that certain combinations worked. Others involved in Modern art were among the first to put parts back together in a creative, usable, acceptable way to consumers, from various geographic areas: Paul Klee (1879-1940); Gino Severini (1883-1966); Fernand Leger (1881-1955); Wassily Kandinsky (1866-1944); John Miro (1893-1938); Salvador Dali (1904-1989); and Marcel Duchamp (1887-1968).

A means of helping to understand the relationship of the various arts, the first of which is more practical, the third usually involves more intuition, all of which can incorporate synthesis, is by considering the three clearly, but simply described categories:

1. The practical arts — the arts of function, such as vocational. They liberate usfrom helpless dependence;
2. The liberal arts — the arts of meaning. They liberate us from a sense of context and continuity; and
3. The fine arts — the arts of transcendence. They teach that people can create more than they can comprehend. (Nealon, John G. 1985)

There is some parallel between these categories and Plato's description of the three arts: "one which uses, another which makes, and a third which initiates." (Republic X, p. 601)

Literature and Poetry

In "The Search for Synthesis in Literature and Art," Colley (1990) uses the illustration of an osprey diving for fish to explain the search for synthesis. In the example, she writes that the synthesis is like joining the object and the desire and that synthesis becomes possible only when there is a gap between what is and what is to be. She points out also that when discussing synthesis the emphasis should not be on difference but on the space (distance) between them. Her statements could also be interpreted that if two objects are together (there is no gap) that they are already synthesized and that further synthesis would occur only at a higher level or as a larger object.

In literature the developments are additive as contrasted to science where the developments are normative as advance builds on the previous. A Shakespeare does not presuppose a Dante. In science, Einstein builds on Galileo, Galileo on Euclid, and Euclid on Pythagoras. (Arieti, 1976)

Poetry is a form of literature "in which the sound and meaning of language are combined to create ideas and feelings" (World Book, 1990), a fertile field for use of synthesis. The free verse, an approach developed in France, and expressed by American-French poet Francis Viele-Griffin (1863-1937), led to the Imagism movement (Imagists) in England in 1912 by T. E. Hulme (1883-1917) , F. S. Flint (1885-1960), R. Aldington (1892-1962), Ezra Pound (1885-1972), and T. S. Eliot (1888-1965). Amy Lowell (1874-1925) was a spokesperson for the Imagists. Carl Sandburg (1878-1967) was a product of this group, primarily following T. S. Eliot. Some poets, such as Sandburg, integrated the physical developments of the day into their messages, just as the previous writers were doing with nature (Naturalists).

The concept of having a vision of the whole as well as the end artifact, as an example of synthesis, is represented by Michener (1988), when he has one of his characters Trevor say, "Point I've been trying to make, the real poet has the last line in mind when he writes his first, ..."

Literature in general has been a field in which writers synthesize approaches to relay their messages. Gilfillan (1848) wrote in Tait's Mag. 15, p. 519, "Milton is the synthesist, Dante is the analyst of Hell," as recorded in the Oxford English Dictionary. In previous centuries the synthesists have usually not been held in high regard by critics and, of course, by the analysts. Critical of synthesists, Hamerton (1873) in Thoughts About Art, xii, p.167, wrote "Synthesists find continual pleasure in observing the relations of things, but from their largeness of range they constantly miss minute truths." With an opposite view, Ralph Waldo Emerson (1803-1882) was one of the early writers to highlight the limitations of analysis when he wrote

> " ...the analytic process is cold and bereaving and shall I say it: somewhat mean, as spying... Were not an ode a better form? The poet sees the wholes and avoids analysis... "
>
> *Emerson (1893), p. 13.*

Will Durant (1885-1981), a philosopher by education, became well known for his approach called 'integrated history.' He explained that integrated history was 'to discover and record the economics, political, spiritual, moral, and cultural activities of each civilization, in each age, as interrelated elements in one whole life.' This is clearly synthesis, looking at the whole and relating the parts to the whole.

As members of society have become more involved with physical systems in recent history, in addition to the natural systems that surround us, the combination of natural and unnatural systems provides another dimension for consideration of synthesis. Environmental control usually involves making an unnatural system more compatible with the natural system. The opportunity for synthesis of new approaches and linkages develops with the additional or new knowledge from science, technology, and humanity related to these systems. No chronicler has received general support for a term to identify the present period in literature beyond Modern (following the Victorian), although there are numerous subdivisions, such as nuclear, age or family, self-focused or 'me,' acquisitive, and for those in manufacturing, an integrated approach. A feature of literature is that it is evaluated on the basis of intense analysis by people from all fields, including the marketplace. As a result of the analysis a new synthesis occurs to initiate, sustain, modify, or terminate a movement.

In 1973 one of the ten best fiction books was "Gravity's Rainbow," by Thomas Pynchon. (Time, 1973) This complicated, massive book represents a synthesis of the natural and the artificial, the spiritual and psychological, the real and the imaginary, and many more contrasts. Devices are used by the writer to create images of destruction, often without considering the constructive aspects, reinforcing a paranoia theme throughout the book. The strength of the book is in the author's ability to bring practically all fields of knowledge and endeavors into a synthesis in which many linkages are developed. These linkages are often difficult to identify and understand at their first introduction of the topic in the book. He incorporates cause-and-effect, phenomenological approaches, system control and out-of-control—terminology often applied to synthetic systems. A lesson conveyed is the need to consider the use and user of technology as it is developed, and sometimes the difficulty in understanding the book is parallel to the difficulty of evaluating a synthesis or synthetic system. His writing is a representation of his statement, 'everything is connected.'

Hayle has gained recognition as a writer who descriptively and convincingly synthesizes literature and science and technology. Her book "The Cosmic Web," 1984 was followed by "Chaos Bound," 1989. The first uses scientific models and literary strategies in the 20th Century. The second suggests that chaos models offer an opportunity to help social scientists conceptualize more reliable means of predicting some human endeavors. The title also includes the contrasting words "orderly disorder' in contemporary literature and sciences. Whereas Pynchon brings in the synthesis of all fields into his writing, Hayle is an excellent writer in a more narrow approach to the synthesis of literature and science. She has devoted a chapter of her first book listed above to Pynchon's work, a chapter that helps understand the work of both writers to a greater depth.

For lack of a better descriptor, Modern has been used to represent the present era in literature, poetry, and humanism. The descriptor Modern will undoubtedly be replaced by another name as the benefit of hindsight is available a century from now. Begiebing (1989) refers to the Post-Modern period as beginning in the sixties. Those who lived through this period in the Western world observed that the sixties stimulated a new sense of synthesis in terms of holistic approaches, interdependence of technology, environment, energy and human systems, supporting Begiebing's views. The important role of synthesis is coming to the fore in literature as represented by two new books:

> Colley, Ann C. 1990. The Search for Synthesis in Literature and Art. The University of Georgia Press. Athens, GA (See introductory para graph).
>
> Begiebing, Robert J. 1989. Toward a New Synthesis: John Fowles (1926-), John Gardner (1933-1982), Norman Mailer (1923-). UMI Research Press, Ann Arbor, MI

In "Toward a New Synthesis", Begiebing (1989) writes about the approach of Fowles, Gardner, and Mailer in their literature since the 1960s. Post-Modernism has had a predominance of "the hermetic" temperament in which imagination may have prevailed but often without relating to the real world. These three authors incorporate issues of the real world, 'including moral dilemmas and struggles in the political, professional, and life conduct.'

"Each of these authors uses a similar structural principle: the dialectical narrative of a hero's or narrator's encounters with the magus figure. Such narrative expresses the dialect of quest, allegory, and romance as psychomania-the debate and flight for mind or soul ...At times, authors who concern themselves with the ends of art rather than the means have led the way toward preachiness, authoritarianism or outright repression." This statement might be interpreted to be in disagreement with a more generally expressed view that in 'synthesis' the end or results should be kept in mind. Rather the statement should be interpreted that the ends and means should be considered together.

The synthesis of these authors is to regenerate classical, Renaissance, romantic, and existential literary traditions, but also to integrate some of the most productive affirmations and scepticisms of Modernism and Post-Modernism. They long to bridge the gaps and discontinuities of life. They long to restore significance to the broken world."

The author points out that Italo Calvino (1923-1985) is a writer who appears to go beyond realism and irrealism and formalism, and perhaps it will take an international author (such as Calvino) to develop a synthesis of more international scope. When we read Calvino we "sympathize, think, and judge." (p. 131, 132)

Some views regarding the literature of the 20th Century, particularly the latter half of the century, are: (EB. 1987, v. 23, 86-224)

» The modern dramatist is no longer bound by theory and works without artificial limits to his work (p. 158)

» Political, economic, and social thought has attempted to reach scientific precision (use of quantitative data, processing devices, and math ematical formulas) (p. 191)

» A condensed or enigmatic sort of prose became preferred to poetry by several poets and the use of aphorism (p. 193)

» The evaluation of fiction has attained the dignity previously held by poetry and drama (p. 162)

» Most of the issues debated in this century appear to be strictly "empirical, even technical, in nature" (p. 206)

» Today it may be almost impossible to come to grips with poetic truth without also coming to terms with positivism (which encouraged a scrupulous regard for the facts) as a total epistemology" (p. 206)

The outstanding trait of 20th Century criticism is the influence of science, the scientific authority and method including the tyranny of analysis, so that the intuitive approach is diminished, and perhaps creativity dulled. In spite of the schools of Imagism, Futurism, Dadaism, and Surrealism, there is no clear summary of the literature "except in terms of its restless multiplicity and factionalism." The author asks, "Would a synthesis of all these methods yield a total theory of literature?" (p. 206)

Western literature appears to have suffered from the influence of the scientific environment which surrounded the writers of the century. At the same time one could wonder whether engineering has suffered similarly. Undue emphasis on analysis by the critics dampened the imagination and the use of intuition by writers (authors). Were the writers in Arabic, Indian, and Oriental, where science is less dominant, more creative in their literature? One would expect that the literature of the Western world would be enriched, enhanced as more sources of scientific information are available to make a synthesis richer and more productive. But apparently that enriched synthesis hasn't happened, except in rare cases as with writers such as Robert Frost (1875-1963) and Carl Sandburg (1878-1967).

Literature, and the other arts, can provide the inspiration for scientific and technological investigations of which recent examples are:

» The possibility that the Shroud of Turin was an artifact of Jesus.

A linen cloth that contained the image of Jesus Christ was claimed to be a burial shroud. Radiocarbon dating showed that the shroud was woven in the 14th Century (about 1357) and could not have been from the time of Jesus. (Vaughn, 1988)

» The claim by some that the 17th Earl of Oxford (Edward de Vere) (1550-1604) wrote many of the words attributed to William Shakespeare (1564-1616). Several other people have been touted as the authors of the works of Shakespeare, such as Lord Francis Bacon (1561-1626), William Stanley (c.1561-1642), and Christopher Marlowe (1564-1593). Using computer models and computer analysis to compare word frequency, often-used words, words used to begin and end lines, use of 'which' and that' clauses, compound and hyphenated words, and frequency of explanation marks, it was found that several works in question were written by William Shakespeare. The analysis and synthesis done by Ward Elliott also found eight poems written by William Shakespeare but not attributed to him, an unexpected synergistic finding. (Miller, 1990) An interesting finding is that others identified did not write the works, but that Shakespeare did write the works, plus others. Without the computer to assist in the study, the second finding would probably not have supported such a conclusive result.

» The possibility that President William Henry Harrison (1773-1841) was poisoned based on a novel. His grave was exhumed, a chemical analysis made' and no evidence of poisoning was found.

» The possibility that President Abraham Lincoln (1809-1865) had Marfan syndrome, a genetic disorder that leads to elongated limbs and a gaunt and gangly appearance. The study was suggested as a result of the possibility that DNA fingerprinting technique of analyzing genes taken from his hair, bones and blood that were saved from the autopsy following his assassination. The study has not been done (1991). (Jerome, 1991)

» Bones found in a Russian forest were identified as those of Czar Nicholas II, his wife and three daughters using DNA analysis, answering long-standing questions about their disappearance.

» The science fiction literature promotes imagination of what might be, often an imagination based on synthesis of ideas from many sources.

Some writers question whether studies should be made that might shatter religious beliefs or myths, that the results could change historical writings of long standing, and findings that expose health data about deceased individuals. The possibility of such forensic studies act as a deterrent to those who would propose or support untruths in their writings. History is better served if it is accurate as possible. Those who favor the study of Lincoln argue that the 'stigma attached to genetic diseases' would be reduced or removed. (Jerome, R. 1991) There is often an attempt to discredit those who have received recognition for their work. That attempt, often with selfish motivation, should be thwarted by providing the correct information where available or attainable.

Music

The epitome of the act of synthesis is the composition of music. The demonstration of synthesis is the presentation of music, whether by an individual or a group, such as a band, orchestra, symphony, or chorus. Presentations of talented composers of previous centuries continue to delight audiences. The 'theses' are the music notes written or scored on a staff with numerous modifications and instructions provided. The 'thing' provided is the written document and the presentation that it represents. One measure of the success of synthesis is the acceptance of the 'user,' the listener in the case of music, or the performer, who literally uses the results of the previous synthesis. Only a limited number of people use the music score as the major product of the work of composers. The final product is usually a presentation to a mass audience.

Early music often had a purpose beyond entertaining. Native Americans usually sang and danced so that the Great Spirit would bring success in personal life, prowess in war and hunting, and provide favorable weather conditions. Early music was often combined with words. The first recorded opera was Eurdice in 1600, and the opera continues to enlighten and entertain audiences throughout the world, often in words not understood by the listener, with the music providing the common language.

Whereas the philosopher is interested in ethical results, the musician is interested in creating aesthetic results. A sort of cause-and-effect is involved in music as the value of music can be measured by its effects. Platonic writings particularly cover a synthesis of music idealogy; Leibniz represented music as a harmony of numbers; and Rosseau wrote that art is an imitation of nature and that music is a copy of the synthesis permeating the universe. (Portnoy, 1954)

The 20th Century marked a drastic change from the music of the past, "as drastic a musical revolution as that of any period in history," (EB. 1961. v. 16, p. 16) A part of music is dependent upon "either conscious or intuitive analysis" *Ibid.* p. 17. The experimentation and variety of approaches defies simple classification, but is often called the Modern or Ultramodern.

Jean Fourier (1768-1830), of heat transfer fame in the engineering world, explained that complicated periodic vibrations may be analyzed into components that have simple harmonic vibrations of single frequencies. Mathematicians have been natural allies of music. Music is not mathematics by itself but is mathematical in structure. The reverse of analysis is the synthesis of musical sound as demonstrated by Hermann Ludwig Helmholtz (1821-1894) using vibrating tuning forks. Today when people consider synthesized music they are usually referring to that sound produced by electronic instrument or controlled by electronic or electrical means and most recently by a computer.

Alexander N. Scriabin (1872-1915) is recognized as an outstanding musician who synthesized music, theatre, art and philosophy using several methods and media in plays, poems, symphanies and sonatos, particularly with the piano. His list of citations is about 300 entries. Another prolific synthesist is Olivier Messiaen (1908-) with 200 entries to his credit.

In 1924 Leo Theremin (1896-1993) invented one of the earliest electronic music devices using electronic tubes, appropriately called the theremin. The theremin was based on use of interference beats of two oscillators to produce sine wave responses and is considered as an early synthesizer. A synthesizer can be used to invent many sounds through creation and combination. One of the first electronic sound synthesizers was developed by acoustical engineers Harry Olson and Herbert Belar in 1955. (EB 8:444) Some synthesizers used additive synthesis in which wanted components for producing music were added, as was done by James Beauchamp in 1962. (EB 24:603) Some sound synthesizers used subtraction synthesis, in which unwanted components were removed from the signal, as was done by Robert A. Moog in 1964, who made the first commercially successful digital synthesizer. The synthesizer can reproduce sounds of many kinds of humans and musical instruments, but can also produce original sounds. The digital synthesizer, which was an electronic device to produce sounds coupled with a computer to control the sounds was developed in 1970s. These new devices revolutionized the type of music, presentation (amplification), participation (masses of people), and the industry associated with it. As a result of synthesizers, the ability to invent (create) has been increased manyfold, and music has numerous variations based on the devices. With the synthesizers the same people are inventing (creating) and presenting (performing) their work (music) to a greater extent than in the past. Formerly, the public performers were more likely to be interpreting and presenting the work of others. Thus, the creativity of the present-day musicians is enhanced with the availability of synthesizers. Most Modern music, from Broadway shows to television and video, depend in unprecedented ways on technology and visualization, tools of synthesis.

SCIENCES

The sciences discussed below are representative. Many disciplines use the word 'science' or add the word 'science' to the name of their discipline, e. g., mathematical science, economic science, domestic science, etc. This is done either to force a greater discipline in their field or to gain the mantle of respect usually afforded to science. There is usually a demand for quality in scientific work, yet it is difficult to describe quality that broadly covers so many fields. The concise statement by John Passmore is that the chief measures of the quality of scientific work is fecundity (richness of imagination) and synthesis (whole), yet a major criticism is that scientists have become too reductionist in their approach. Insiders in science tend to use quantitative measures such as numbers of papers published; outsiders tend to use more qualitative measures in addition to numbers to learn the impact of the work.

The nexus of philosophy, arts and sciences covered in this part is extended by the nexus to technology and the professions discussed in following parts. (Image 5-3)

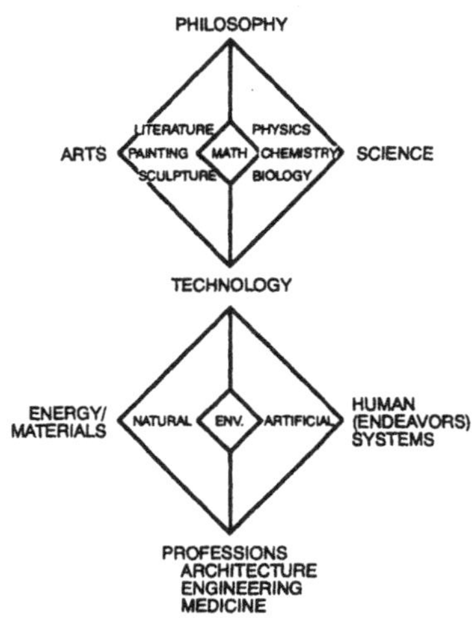

Image 5-3. Philosophy-Technology-Professions (Cassidy, 1962)

Biology

In biology in the mid-19th Century the approach of biologists was reductionism. (Tauber, 1991). Although many advances resulted from that approach, a lack of considering the organism as a whole limited major advances. A holistic approach was then taken by opposing groups. In the early twentieth century an integrative approach developed, in which people focused on the interrelationship of parts. (Allen, 1978). A parallel development occurred in evolutionary biology resulting in a synthesis, the evolutionary synthesis, of population and Mendelian genetics with Darwin theory. (Mayr and Provine, 1980) (Tauber, 1991). The synthetic theory of evolution was developed in the 1930's and 1940s lead by T. G. Dobzhansky (1900-1975), E. Mayr (1904-), and Julian Huxley (1887-1975). The theory kept changing as new discoveries including molecular biology and DNA were made, and work of the chemists and physicists was incorporated.

Jean Louis Agassiz's (1807-1873) approach in biology of observing facts and their orderly arrangement was accompanied by the exhortation not to be content with them. 'Facts are stupid things until brought into connection with some general law,' referring to the importance of going beyond analysis and to incorporate synthesis. (Lee, 1954)

The discovery of the double-helix arrangement for the DNA molecule was a remarkable finding through intuition by Francis H. C. Crick and James D. Watson. They used the data of others and conducted few studies on their own. A short two-page article in Nature 171, 737-738, April 25, 1953, was the basis of their (along with Maurice H. F. Wilkins) 1962 Nobel prize in physiology or medicine, one of an increasing number of Nobel prizes to go to synthesists in contrast to analysts. The DNA molecule was discovered much earlier, in 1868, by German biochemist Friedrich Miescher, but a description of its configuration was not settled until 1953. In 1972 the first DNA molecule was cloned and by 1977 the methods for rapidly sequencing the DNA molecule were developed. And only five years later genetically-engineered insulin was available for treating diabetes. Biotechnology advances will contribute to improved health, plant agriculture, food with higher nutrient content, crops with better yield and higher resistance to disease and drouth, environmentally benign manufacturing operations, bioremediation, and manufacture of new products being developed based on the DNA knowledge. The human DNA molecule is now being used as a sensor to monitor environmental effects on the body. A device that contains the human DNAmolecule is being developed that will measure the surface exposure to ultraviolet effect as a result of ozone depletion in the atmosphere.

Some biologists today are seeking for a new synthesis to find a conceptual format to which the organism becomes the focus rather than the components. (Tauber, 1991). Biophysicists and biochemists, people prepared in two fields, emerged in the mid-fifties to meet the special challenges for integrated or interrelated approaches. It seems however, that systematic biology is a field in decline (in England) as proposals on systematics submitted to research councils for grants do not fare well, as the peer review process favors analysis types of projects (Gee, 1992). This criticism is mirrored by many others who propose synthesis types of projects, not only in England, but in the United States.

Developments in biology, particularly in molecular biology, have led to many advancements in biotechnology and bioengineering. These developments include knowledge of subcellular units (including RNA and DNA), microbial diversity (use of oxygen, temperature effects, pH tolerance), cellular physiology (growth characteristics of whole and combinations of cells), and genetic engineering (insertion of foreign substance, such as DNA, into the cells. (Myers, 1987) Parallel to these biological advancements in knowledge has been the equally important instruments and equipment that made possible some of the findings.

A recognition of the importance of synthesis in biology is not new. Biology itself in the natural world is an activity in synthesis. Most recently new academic units have been formed to recognize these advances, the latest of which is a Department of Integrative Biology, University of California, Berkeley, and other universities that have established units, sections, programs, or groups for the subject.

Biodiversity, biological diversity, has been receiving much attention for biological systems from the level of species to ecosystems. A variety of plants and animals living in various environments helps assure a longer life, greater overall productivity, and an ability to withstand possible domination by a particular species or variety. Biodiversity provides a short term dynamism and a long term stability to the system. The principle might well be applied to other systems—political, economical, and perhaps, fuzzy mathematics.

The idea of acquired characteristics being transmitted to offspring has been strongly discouraged, as represented by the villification of Lysenko, and by opinion surveys of biologists, and I presume many letters to the editors. Whereas biologists have almost lock-step opposed thinking along the line of transmitting acquired characteristics by heredity, there is now a willing to discuss the possibilities. Landman (1993) reports a substantial number of experimental systems that display inheritance of acquired characteristics, such as caused by a brief change in the environment, and in which the new traits in treated animals or cells persist when put back into the original environment. A phenomenology approach by sudden or induced massive change can occur. The mechanisms involved are compatible with modern molecular biology. It would be interesting to know how many potentially outstanding biologists have suffered and new knowledge rejected as a result of the restrictive approach. It may well turn out that those who vilified Lysenko for being politically correct (PC) were really the ones who were being politically correct (PC) and based their thinking on limited information.

Chemistry

The 20th Century inherited a field of chemistry that was primarily analytical, often qualitative, and mostly inorganic. Dmitri Mendeleev (1834-1907) (1869) discovered the periodic law of elements which was published in 1869 and invented the table representing the relationship of elements, based on 63 known elements at the time. His table was one of the great organizing principles of nature and was used as a predictive tool for existing but undiscovered elements. The table was the major contribution in synthesis, a synthesis of information to form a more coherent system, not the usual synthesis associated with assembling chemical compounds. Rapid developments occurred in organic chemistry, quantitative chemistry, and physical chemistry. Biochemistry, often nurtured in chemistry departments, split off as a separate discipline. Many of the advancements were spurred by new precise, accurate instruments, the availability of which provided the basis for a new generation of instruments and a need for an increasing willingness for people in different fields to work together. The new instruments coupled with computer systems revolutionized the analysis and greatly increased productivity and made it easier to incorporate the processes into a commercial operation.

The first synthetic material was a product of Frenchman Braconnot, who in 1832 accidently poured nitric acid on cotton (cellulose) and then on wood to produce an impermeable film. That discovery was followed by Swiss Schoenbein who developed the first transparent paper, cellophane, in 1840 by accident.

In the thirties commercial attempts were made to produce substitute materials, such as using agricultural biological products to replace petroleum based items and crops to replace rubber, as represented by a field called chemurgy. In 1939, John Wiley & Sons publishers inaugurated a Synthetic Inorganic Chemistry series, and in 1941 a Synthetic Organic Series of books. An international journal on methods of synthetic organic chemistry, Synthesis, was inaugurated in 1969 and continues with twelve issues per year. In 1971 two other journals began, Synthesis and Reactivity in Inorganic and Metal-Organic Chemistry, and Synthetic Communications (in synthetic organic chemistry). Synthesis of materials, to replace other materials, and the invention of new materials, has always been a part of chemistry.

In the last twenty five years new departments or sections of Synthetic Chemistry have been formed in universities and industries. These departments were often formed outside the conventional Chemistry Department and, as in Japan in some universities, outside of the College of Sciences that usually houses the Chemistry Department. This happened because some people believe that the synthesis approach is not compatible with the analytic approach so strongly ingrained in the personnel who inhabit many Chemistry Departments. Thus, a different institutional arrangement was required which probably resulted in better funding.

Using the computer along with other analytic tools, a chemist can perform retro-synthesis or reverse synthesis of a substance working back to the basic starting materials to find optimum pathways to facilitate the pathways to efficient processes. This is similar to reverse engineering, so well known in industry. The 1990 Nobel prize for chemistry went to Harvard University researcher E. J. Corey who developed an ingenious approach to synthesizing complicated molecules that has revolutionized the development of new drugs. The interrelation is illustrated by an article on computer-assisted analysis in organic synthesis. (Corey, 1985)

Animal and plant breeding has been subject to human manipulation since the time of Mendel, and probably before his time, to synthesize a new or improved product. Interference in the natural breeding process can result in products with different characteristics than would otherwise occur. Who can say that the 1.6 billion cattle in the world are natural because many may be unnatural in the sense that they resulted from human-controlled breeding, called artificial breeding, but we don't call these artificial cows.

The field of materials has become a more visible area of investigation for chemists. Usually, synthesis is involved to "engineer" a material with specified characteristics, properties, or uses. One of these of interest is the newly discovered C_{60}, C_{70} or more carbon atoms (1985), buckminsterfullerene (buckyballs), a form of carbon formed when vaporized carbon condenses, whose properties are now being determined. The name is based on R. Buckminster Fuller, who was hardly a chemist, who developed the geodesic dome that is similar in shape to the C_{60+n} atom. The material has potential to be used as an electronic material, high temperature superconductor, magnetic material, catalyst, microscopic capsules, and filter. The buckyballs can be used as a cage in which various elements or compounds can be placed to make a chemical tracer; to encapsulate medicine or radioactive atoms for medical uses. The buckyballs can be used as a matrix to which atoms or elements or molecules might be attached. Synthesis will continue to grow on a large scale during the next century as an approach of both academia and industry.

One of the most creative chemists of our time is Raold Hoffmann who combines analysis and synthesis in his approaches. He writes of the division between natural and unnatural means of the sources of chemicals. By doing so he avoids the confusion sometimes engendered when using the terms "synthetic" and "artificial" and "imitation," which can be incorporated in the term "unnatural." Synthesis may be done by mimicking nature or by using an entirely different process, either suggested by observing the process of nature or by an entirely different approach resulting from the creativity of the person. During the past 300 years a majority of the unnaturally formed chemicals were formed providing some 15 million well-characterized molecules. Also, one might argue that Hoffmann's interest in poetry and art has enlarged his vision, creativity, and productivity, or vice-versa, that his prodigious production in chemistry has enhanced the arts. (Hoffmann, 1993)

Mathematics

In the early years mathematics was a field associated with philosophy and much of the early writings in philosophy were done by mathematicians. As philosophy developed and the arts category was recognized, mathematics was included in the arts. Mathematical propositions are judgments *a priori*, based on prior knowledge and noted for the beauty of the solutions to problems, not empirical and not necessarily practical. Mathematical reasoning is deductive. Newman, in his History of Mathematics, states that "Intuition fails in the development of both elementary and complex mathematics." That does not mean that the development of mathematics by intuition is wrong. The 20th Century, particularly the last half, has seen an order of magnitude increase in the use of intuition in mathematics. Casti (1990) expresses the view that "mathematics is intuitive... justified by proof."

In the 20th Century, mathematicians have given more attention to the design of experiments and interpretation of results of experiments. Statistics and probability, examples of applied mathematics, have often developed outside of the academic mathematics. Statistics as an applied mathematics led in developing theories in the field. Statistics became a major tool for analysis by empiricists and experimentalists. Statistics provided a procedure for helping to make judgments regarding functional relationships involving a very large number of data by using sampling procedures and using limited calculations. Now the computer is available that can handle large numbers of data (all data can be fed into the computer quickly by scanning devices) and can make complicated calculations quickly, a new approach, a new statistics probably not called statistics, is in order.

Much of the research in science fields consists of evaluating one variable. Such research can be the basis of an article. A limited number of variables, one or two, are often treated in a science article. In engineering and other applied fields, working in the 'real world' with several interacting variables, requiring judgement on the relationship of the variables with the phenomenon. The results of analysis serve as a guide, but the variable showing the greatest effect on the phenomena may not be the most important one when several interacting variables are considered. Response surface methodology (RSM), a statistical approach provides a means of guiding one's research and practice to find the relationship of responses with several variables.

The AIDS (acquired immune deficiency syndrome) epidemic highlights a need for new approaches in statistics and experimentation. Trials on the effectiveness of certain drugs that appear to have the potential of assisting the patient involve use of a placebo for the consumers being tested. If it turns out that the drug being evaluated has favorable effects, perhaps even life-saving, those people taking the placebo are potentially deprived of the benefits of the drug.

Is there another approach?

Applied mathematics took on new meaning during the 20th Century.

Operations research and systems engineering were developed using the tools of mathematics. These fields became prominent as people began to develop and use a synthesis approach while keeping a quantitative approach. To adapt these quantitative tools to systems with numerous variables, some of which were not separately quantified, fuzzy logic, fuzzy systems, and lumped parameters were developed. Fuzzy logic evolved from work by L. A. Zadeh in 1965 on fuzzy sets. A procedure has evolved in which a control can be on the basis of qualitative terms such as cold, warm, and hot. Fuzzy logic is a powerful tool for engineers and deviates from the usual thinking of mathematicians and also applies to some other imprecise applications in medicine, meteorology, and psychology.

Many fields of science, engineering, and technology discovered the value of quantification, had new tools and instruments of measurement available, and during the last half of the century, acquired computer hardware and software for quick, accurate, and complicated calculations. Thus, discrete analysis and finite analysis became important subjects often developed outside academic departments of mathematics. In applied mathematics, the mathematician would focus on the basis of the calculations of an experiment, not so much on the results obtained by the experimenter. Interest has increased in the mathematics of discrete functions, particularly applicable to digital systems. But in the future an increased synthesis by the mathematician and experimenter will emerge.

The emergence of computers has revived some of the fields of mathematics considered as pure, such that Boolean algebra and Lie groups, became applied and provided a driving force for research in these and other areas.

With the advent of the computer mathematicians have tackled challenging problems, often stated hundreds of years ago, but considered unsolvable since then. Recently, Fermat's Last Theorem, proposed about 1637, believed to be true by most mathematicians, but was not proven to be true until 1993 by Andrew Wiles of Princeton University. (Cipra, 1993) (Stewart, 1993)

The procedure for multiplying, dividing and square rooting Roman numbers Roman numbers went unanswered until solved by a high school student, French Anderson, Tulsa Oklahoma Central High School, not with a computer but by creative effort. (Time, 1956)

In recent years mathematics as a field has been included with the academic sciences. This move of mathematics from the arts has resulted because of the mathematicians use of logic, quantification, and use of mathematics in other fields of science and technology, to a greater extent than its use in the arts. The availability of the computer has provided an opportunity for the use of mathematics (applied mathematics) often not welcomed by the purists or theorists. Mathematics can be considered as analytical (differential) and synthetical (integral), another justification for juxtaposing it between the arts and the sciences. There is no question that the techniques for synthesis will be enlarged with the increased scale and number of components and systemization of natural and artificial systems.

Physics

Physics changed dramatically in the 20th Century, primarily due to the revolutionary concept of relativity proposed by Albert Einstein. Einstein developed many theories based on intuition, but at the same time couldn't accept during his lifetime by intuition the quantum relationships that he helped develop. Disagreement continues on the propriety of quantum theory as espoused today. Closely allied to philosophy in the formative years, then becoming the vanguard in the category of natural philosophy, physics has attracted leading thinkers and produced prodigious findings. J. Larmor in his introduction to Poincare (1902)(in 1952 translation), recognizing the limitations of excessive analysis, wrote," ...until our apparent continuities are resolved mentally into discrete aggregates which we only partially apprehend. . .The irresistible impulse to atomize everything thus proves to be not merely a disease of the physicist; a deeper origin, in the nature of knowledge itself, is suggested."

Although much of the early work was based on analysis, the 20th Century physicist has played a leading role in synthesis, synthesis of the available and new information in matter and materials, electricity and electronics, gravity and magnetism, and radiation and quantum. The knowledge of the nucleus became the basis of developments in atomic and nuclear energy, such that this century may become known as the beginning of the Nuclear Era. The idea of a unification principle has challenged physicists to develop comprehensive models of the universe. (Image 5-4)

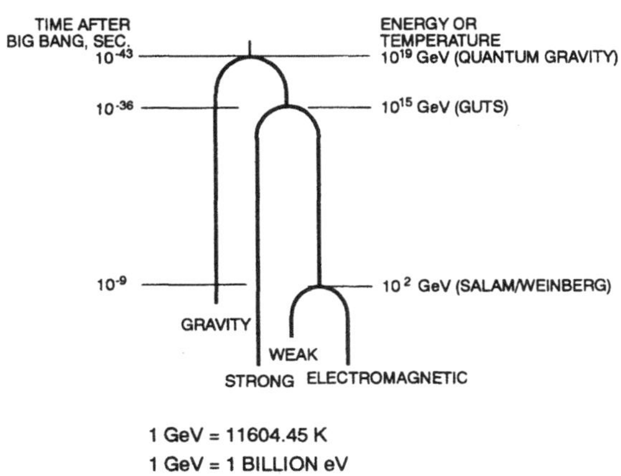

Image 5-4. Unification Scheme of the Universe (Scott, 1991)

Physicists began to look at large systems and identify voids in the information to be used as the basis of further research. Physicists speculated about natural systems, much as Copernicus had done in earlier years. And near the end of the century that speculation lead to advances beyond the earth in black holes, quasars, and planets and galaxies beyond the Milky Way.

Although analysis is usually considered the core of the scientific method, the composition of water was originally discovered by synthetical and not analytical processes. (Huxley, T. 1877) Amedeo Avogadro (1776-1856) predicted the number of molecules in one mole or gram-molecular weight of a substance. Sixteen independent approaches were taken by investigators to verify the number, resulting in agreement by connectivity, a feature of synthesis to obtain a basic physical constant. (Morowitz, 1993) The value of Avogadro's number was established as 6.02×10^{23} Isaac Newton (1642-1727) first analyzed daylight using a prism to identify the various colors of the spectrum. When he resynthesized the various colors of the spectrum he discovered that he could reconstruct the daylight by using only three of the elementary colors. The three color principle is used commercially in coloring prints and mixing paints. This synthesis is similar to synthetic chemistry from which many new and useful dyes and textiles have been created.

In the 17th Century Galileo Galilei (1564-1642) was charged with heresy by the Inquisition in 1633 for his view that the sun not the earth was the center of the universe. Such views were not religiously correct (akin to today's politically correct). It was not until 1984 that the Vatican released the early records and cleared Galileo, saying that the Inquisition by the church may have erred.

Kepler searched for a unifying picture of the world at the time that there were two camps, one claiming that the earth was the center of the universe, and the other that the sun was the center of the universe. The first was based on the claims of Ptolemy (c. 75 AD) as a geocentric world, and the second was based on the claims of Copernicus (1473-1543) and strengthened by Galileo (1564-1642) as a heliocentric world. Kepler (1571-1630) was a comprehensive thinker (synthesist) who met Tycho Brahe (1546-1601) (analyst) from whom he learned the importance of precise observation and analysis. Holton (1988) writes, "For without some such support for the imagination, coming to us from beyond the boundaries of science alone, without the help of all the best that has been thought and felt before us, how could we hope that the brief attention we can give to scientific problems during our short lifetimes could ever yield anything worthwhile?"

Many physicists now work as members of teams or groups. Since World War II there has been a trend toward large comprehensive projects, many international in scope, inspired by the Manhattan project, and the number of physicists working alone has decreased. The core of advanced physics has moved toward high energy studies with large facilities and expensive equipment, called "Big Science." The magnitude of major projects often requires international funding and numerous personnel. A recent article on electromagnetic processes and properties published in Europhysics Letters has 173 authors whose listing occupies 29 lines. Presumably all of these people were involved in doing the comprehensive research at European Center for Nuclear Research (CERN). (Arnison, G. ... Zotto, P. 1986)

Major projects that require and merit international cooperation include those dealing with the environment, particularly where the decision of one country impacts the environment of another country (stack gases from U. S. flowing into Canada), space travel to other planets (space stations for human occupancy), pollution that affects the atmosphere (chlorine compounds that decrease the protective ozone layer), and mitigation against natural hazards that span the earth (earthquakes, tsunamis, violent storms and volcanoes). Incentives and recognition are required to motivate people to prepare for, to participate in, and to produce under conditions that encourage international cooperation.

ADULATION OF SCIENCE

The adulation of science-as represented by adding the terminology of science to conventional activities to try to make respectable, even in advertisements, causes confusion and misunderstanding by the public. The conflicting verbiage in which engineering science used as engineering adds to the misunderstanding. Engineering differs from science in its objectives and its methods of operation. Science-scientists can reach substantial agreement in results by adding more, well-prepared people in scientific work, progress is made more rapidly in a creative environment. The production of technological devices based on science to provide benefits to people has become a public enterprise and thus commands public attention. The over adulation of science and unusual unfilled expectations could lead to its 'downfall.'

SUMMARY

The antagonism between the two cultures of the arts and the sciences as expressed by C. P. Snow continues. One hope for decreasing the divisiveness of that situation is the trend toward increased synthesis in thinking and in action, not only within the well entrenched disciplines, but across disciplines and activities. The continued emphasis on reductionism will not ameliorate or heal the breaks between the arts and sciences, but when combined with deduction and induction will help establish a common ground for reducing the tensions. The two modes of thought, empirical-inductive and theoretical-deductive, discussed by Conant (1964) applied primarily to science, and does not, nor is it intended, to apply to the broader field of sciences and arts. Analysis-synthesis as in exercise in unity offers hope for bringing the arts and sciences groups closer together. The concern for human welfare, energy, and the environment, will be a driving force to consider not only natural (including biological) and physical systems, but social and human systems as well. Interdisciplinarity and multidisciplinarity will take on new meaning. The educational system must incorporate the synthesis approach as a means of accomplishing holistic objectives. As much scientific research is funded from public sources, members of society must be kept informed of the activities and results and potential impact on society.

Part 6.
Trends in Synthesis in the Learned Professions and Applied Fields

THE LEARNED PROFESSIONS

In the learned professions the practitioners build on a scholarly base, gain additional learning by experience, and relate to serving the public directly. They have the freedom and responsibility for individual action, adhere to a code of ethics and are licensed or certified by society through government to practice. They are classified in four learned professional groups—architecture, including landscape architecture and interior design; engineering, including surveying; law; and medicine, including dentistry, optometry, and veterinary. They will be considered for this presentation by focusing on the synthesis aspect of their activities. These professions feed upon the arts and sciences and provide feedback to the arts and sciences. Many other groups require approval of some type to advertise and practice their profession with the public, such as meteorology and nutrition. In addition, some observations will be made regarding agriculture, business, education, and some of the social sciences. These professions, along with the scholarly professions (perhaps these should be called fields), usually require a college degree or equivalent as compared to the skilled professions, for which intense training is needed.

The early and often identified as the original professions were law, religion, and medicine. Architecture and engineering followed. During the 20th Century many groups unrelated to the original professions, groups seeking to associate with the positive image of the professions, have become self-proclaimed professions. Some people both inside and outside the professions have tended to diminish the favorable views of others of the professions in the past. To the extent that the professions become self-serving they risk losing the confidence of the public.

The principal professions as now practiced are classified as follows:

» scholarly—philosophy, religion, etc.

» learned—architecture, engineering including computer science and technology, law, medicine, etc.

» skilled—athletics, coiffeur, culinary, police, etc.

Whereas the first category is strongly knowledge-based, the third is strongly skill-based, and with the second, the learned professions, being a blend of knowledge and skill.

Architecture

Until the 1970s the architecture of the middle of the 20th Century was identified as Modernism. This was followed by a period of Post-Modernism developed as a reaction against modernistic architecture, associated with the work of Robert Venturi and others such as Peter Eisenman, Michael Graves, Charles Moore, Robert Stern, and Stanley Tigerman. (World Book, 1990) They revived some rejected styles, incorporated ornamentation, and borrowed from previous styles, such as arches and columns. The Post-Modernism movement is a synthesis of the past with creative ideas for the future, and as expressed by Venturi a synthesis of structural, spatial, and formal values. Here is a recognition of synthesis as an approach used by architects as a path to a final product.

In the early part of the 20th Century, the Bauhaus school of architecture, founded 75 years ago (1994) encouraged ornamentation as an integral part of the building totality, a kind of synthesis, and that if one could design one item, another item could be designed. Gropius (1935) wrote, "Our ultimate goal ...was the composite but inseparable work of art, the great building, in which the old dividing-line between monumental and decorative elements would have disappeared for ever."

The Bauhaus program, the synthesis of form and function, uniting art, engineering, and craftsmanship founded in 1919, continues to be perceived as modern.

Architecture has always been closely associated with art. The early architects were often artists. Goethe (1749-1832) describes architecture as frozen music. The Encyclopedia Britannica (1987) describes the distinction between the theory of architecture and the theory of art. The ideas of art are based on beauty, esthetics, and form, for example, and are incorporated into architecture. In architecture form and function, esthetics and environmental ideas, and beauty and cost are considered in developing a useful structure. Architects are encouraged to take a larger view of their work, a more holistic approach, which translates into a greater synthesis. Getting the proper balance between analysis and synthesis is a difficult challenge. In some cases the works of architects are criticized because of an apparent lack of concern for details, such as roofs or foundations that leak, but these shortcomings are usually more than compensated by a bold synthesis in which the design considers the occupants and their tasks, congruity of the building contents and function with design, blending of the building and the landscape, and choice of materials, conventional and modern. Architects spend considerable time and effort in analysis, but that effort is usually for the purpose of improving the synthesis. That trend is expected to continue, and new names of people and trends may emerge to more strongly identify that approach.

Education in architecture emphasizes the importance and role of the client, perhaps to a greater extent at the early stages than the medical profession. Juries of student work include practicing architects who provide critiques that initiate the student at an early period in the profession. In addition, seminars, shadow programs, adjunct professorships of practitioners, and internships expose the students to the profession, so the importance of the user becomes well-imbedded.

Some Schools of Architecture grew out of engineering. Some schools offer Architectural Engineering, a combination of the two areas, the latter emphasizing engineering. Architecture and Engineering remain closely linked in their activities although the close linkages in practice often becomes antagonistic. They are sometimes classified as design professions, where the overall continuing activity is synthesis. Synthesis requires a visualization of the final product so that the elements or components are appropriately used and assembled. (See Part 7) Leonardo da Vinci is often classified as an architect and as an engineer, and is a famous artist, who had an unique approach to his work, one of seeing the whole and thinking and performing as a synthesist.

Engineering

Engineering, as other learned professions, draws heavily on the arts and sciences. Some people classify engineering as a science, but that classification puts too much emphasis on analysis as a method. At the same time that classification denies engineering of the importance of art in its activities. Both those who are practicing the profession and those who are served fail to see engineering as a human endeavor, except perhaps for those in the biomedical and biological engineering fields. Certainly the aspect of focusing on the end use of a product, system, or service is common with other professions, such as architecture. Engineering is devoted to unnatural manmade, constructed, or manufactured "things" instead of natural "things," the focus of sciences. By providing artifacts that replace or substitute for natural "things" engineering is devoted to synthesis, but it is more than that. Its engineers "think" or imagine, or perhaps more appropriately to image "things" to be made, and then design to make them. And that makes engineering truly a profession of synthesis.

The concept of the role of engineering varies greatly in different countries, possibly related to the cultural underpinnings of synthesis versus analysis. This relationship is illustrated by the notion that countries with the Romance languages are more attuned to art and those with the Germanic languages are more attuned to technology. The Asian culture provides a stronger synthesis while the English culture is more heavily analysis. This attitude is illustrated by observations of Wang (1992) who noted that the "Japanese bioprocess engineering industry does what it needs to do in order to get a product produced in a cost-effective manner; however the Japanese do not seem to overanalyze (and agonize) over the best mode of producing a product, but simply develop an acceptable mode of preparation," and go from there.

The strong ties of engineering to the sciences, particularly the physical sciences, are evident by the large number of people from the sciences who practice engineering, and the large number of engineers who are called scientists. Josiah Gibbs (1839-1903) made important contributions in the field of thermodynamics. His formulation of the phase rule, specifically to the gaseous, liquid, and solid phases of water, but generally the phase rule has been applied to other matter than exists in more than one phase. Gibbs received the first doctorate in engineering in the United States which was given by Yale University in 1863. Interestingly, practically all the literature refers to Gibbs as a scientist.

The Scientific Method developed for natural systems is used as a tool for many engineers working on artificial systems. The Engineering Method more appropriately represents the approach for a majority of engineers. Engineering, particularly for large structures or systems that cannot be evaluated in the laboratory, has much in common with the observational sciences, such as astronomy and oceanography, where the elements of the natural system cannot be manipulated directly.

Engineering and architecture are closely linked in their professional approaches, although professional jealousies seem strong. Both professions are involved in design of a product to meet the need of a client, an industry, or the public. Architecture has been more closely linked to art and beauty than has engineering. In contrast engineering has been more closely related to the physical structure, construction, environmental control, and cost and are often criticized for a lack of beauty in their structures, with the possible exception of bridges.

> "The engineer, inspired by the law of economy and led by mathematical calculation, puts us in accord with the laws of the universe. He achieves harmony. The architect, by his arrangement of forms, achieves an order which is a pure creation of his spirit. . .it is then that we experience beauty."
>
> *[originally in French] (Le Corbusier, 1927)*

Engineering has more to learn from the arts than is usually recognized. Engineering grew out of the crafts (activities in which the design is usually done by others). Engineering has struggled for years to cast that association aside and to replace the crafts (manual skills) with sciences. Some leaders in engineering education, seeing the need have over zealously pressed for a stronger science base. A stronger science base was required, without question. The approach should have been to add the sciences to an updated arts or to synthesize the two, rather than to replace the arts, as occurred in many cases. Engineering as a profession, unlike architecture, has been unwilling to extend the length of education for the student to enter the profession, in spite of the tremendous increase in the knowledge base. Only with a greater synthesis of the knowledge base with engineering objectives can the education of the engineer to enter the profession be done in four years. About half of the engineers obtain an advanced degree but this is after or could be after they have completed the degree for entering practice in the profession. Because a graduate degree (advanced degree) is not required to enter the profession, some people deny the status of engineering as a learned profession.

Poetry and literature require imagination, synthesis of thoughts, need to proceed with incomplete information, consideration of social issues, and putting substance to thoughts through theses (words, treatises, and designs) to provide things. A challenge is to incorporate these attributes into the educational program. One way of accomplishing that objective is to expose engineers to these subjects through their education, both in separate courses (which is highly unlikely) and in engineering courses incorporating these subjects. It has been said, "Engineering has more to learn from poetry than from physics."

Numerous individuals are calling for the education of engineers to include more on synthesis, and to incorporate those subjects that would provide the basis for synthesis of systems that include the biological sciences and social sciences as well as the arts. These thoughts are represented in Part 9 by Bordogna, 1990; Drexler, 1986; Fuller, 1989; Kash, 1989; Kellogg, 1987; and Simon, 1991.

The preoccupation of engineering with analysis is attributed by some to the Taylorism of the early 20th Century. During this period considerable attention was given to analyzing worker's movements by description and timing developed by Lillian and Frank Gilbreth, with the minute elements of activities called therbligs (Gilbreth spelled backwards). There are some writers who argue that minimal attention was given to the human factors by Taylor in conjunction with the therbligs. Those writers did not give proper attention to the writings of Frederick Taylor about management. He wrote of maintaining close personal relationships between management and other workers, strengthening the working group, and developing the individual for the job. Although use of therbligs may have been a factor, the tendency of engineering to copy the methods of science probably had more impact on developing the strong analysis dimension in engineering than the use of therbligs for work study.

From a mechanistic approach (easy to attribute to engineering approach), the reductionist thought suggests that the whole is not greater than the sum of the parts. Perhaps it was better stated by Aristotle, although the father of analysis, who recognized the importance of the 'whole':

> "The whole is more than the sum of the its parts.
> The part is more than a fraction of the whole."

In Engineering, synthesis is the process, incorporating integration, interfacing, and networking, of which the product is the system. Analysis is used as a method of improving the synthesis as a whole and the part in particular. This approach depends more upon inductive, intuitive, and heuristic thinking, an inventive nature, considering the general or universal, and speculative as means to an end as compared to analysis. Yet both are vitally needed. The knowledge base of the engineer constructed on intuitive and heuristic thinking, is called doxa, that is knowledge true beyond a shadow of a doubt, but nevertheless, may be changed.

Law (Jurisprudence)

Jurisprudence is legal philosophy, the philosophy of law. Analytical jurisprudence represents the study of law in terms of its logical structure. The term synthetic jurisprudence is not used in law literature, although the word synthesis as relates to ideas, approaches, and methods is used. Laws are human-formulated in the social world, just as devices and artifacts are human-made in the physical world. The human-made laws fit into and conform with the larger workings of the universe and human nature of the world. These laws are designed (some say crafted) to accomplish certain desired results. The practice of law is primarily argumentative in which the processes of science, deduction and induction, and creativity are used to persuade. However, there are opposing views in the legal profession that could well be categorized under analytic and synthetic. The analytic view is represented by Aristotle, who from a scientific analysis of rational principles, and the synthetic view, at least partially, by Plato who attempted to develop his concept of justice from inspiration. (Friedmann, 1967)

Murphy and Coleman (1984) contrast analytical jurisprudence and normative jurisprudence. The first deals with logical analysis of the basic concepts that arise in law, such as negligence, duty, responsibility, concerned with "as the law is." The second deals with evaluation of legal practices and is more concerned with "law as it might be." Normative means describing, prescribing, or creating a norm, heuristic in nature, often including aesthetic values, and as such are normative approaches to synthesis. (Webster)

Law-making deals more with generalization, but the application of law deals with particular cases.

> "This fundamental philosophical problem of the general and the particular is all but ignored in our texts."
>
> *(Kerruish, 1991)*

The lawyer is more involved in synthesis than is generally recognized by both the scholar and the practitioner. An imaginative, or an imagined activity or event can challenge lawyers attempting to identify the many possibilities of conformity with or conflict with the written law. Mere analysis may not be too useful without reference to a larger body of knowledge or without considering imagined possibilities. So analysis-synthesis are involved in both law-making and law interpretation, and law enforcement, although the terminology in the field is different than in other learned professions.

Society relies on the predictive quality of law, in that the societal good is furthered when one is able to accurately anticipate the legal consequences of his actions. This predictive quality is based on the consistent and uniform interpretation of statutes and regulations and the application of settled legal principals, or precedents, to new facts. This latter quality, known as *stare decisis*, was the foundation of Roman law from which much of the Anglo-American jurisprudence is based.

In this technological age, the practice of law has also become more synthetic and interdisciplinary than at any time in the past. Nowadays, more and more lawyers must draw on knowledge from the sciences, industry, and the other professions in order to capably render legal advice. For example, patent lawyers usually possess technical skills and must continually draw on these skills in their practice. Similarly, environmental lawyers typically need to have a working knowledge of chemistry and environmental engineering concepts in order to apply the increasingly technical environmental regulations in a proper manner.

The legal field needs to be considered very much as the engineering field in terms of principles or theories and practice. For the average practitioner synthesis is an integral part of their everyday work. In dealing with a client, a user so to speak, an attorney is presented with the needs, wants and circumstances of a particular individual or entity. The attorney then must create something in the form of documents, agreements, depositions, estate plans, contracts, and cases to satisfy the client's needs. Although certain elements may be included in many agreements, contracts, and plans, the attorney must synthesize the given data to come up with a suitable result to serve the client.

In a court room a particular result is much less certain where the attorney uses facts, relevant law, and personal skill to obtain favorable results. There are tangibles, such as the facts of the case, relevant laws, rules, and regulations; there are intangibles, the judge, jury, attorneys, and the interpretation of tangibles. The interpretation aspects allows the lawyer to manipulate the facts and laws to lead to the desired result. A synthetic approach is used to create an interpretation desirable to the client. We see many instances when this use of synthesis may benefit a particular individual, but not society as a whole, by encouraging frivolous litigation. Business, estate planning, tax, patent, copyright, and real estate are heavily involved in crafting documents, using more of an analytic approach. Criminal, civil litigation, bankruptcy, personal injury are more heavily involved in using synthesis.

Some people in the legal field use the term 'systematic' to approach what some of us in other fields call "synthetic." Whereas, law and legal books do not refer to synthesis they do use the words "system" and "systematic." Systematic refers to the property of a thing described rather than a common property of its constituent parts. (Sampford, 1989); although he suggests that law is best seen as not systematic, he says that the attempted or forced systemization of nonsystematic phenomena leads to problems. Yet an appreciation of law in general, and specifically whether law is systematic, and therefore whether or not their activities are a part of a larger system is important. (Sampford, 1989, Integration is a step toward synthesis, and the integration of legal activities as a part of the social, political and economic fabric is shown in Image 6-1.

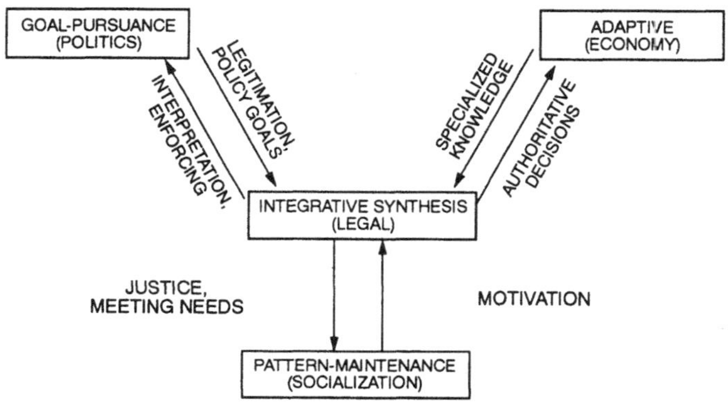

Image 6-1. Horizontal Structural of Society with Legal Basis
(Sampford, 1989)

White (1973) writes of the distinction between the mind that tells a story and the mind that gives reasons: one finds its meaning in imagined experience, the other in systematic exposition of conceptual order or structure. One is given to narrative, the other to structure. The lawyer must do both, the point being that analytic jurisprudence, although the focus of much legal writing does not adequately represent the legal profession's approach. The tensions between story and theory, fact and fiction, and the particular and general are more important than usually recognized.

Lawyers and judges can benefit from the use of general systems theory by viewing a problem or a process in terms of its total relevant environment. An integration of parts of analytical and sociological jurisprudence are well intertwined in law and policy issues. This represents a more synthetic approach. according to Conant. (Dias,1979).

Friedmann (1967) urges lawyers to expand their views and approaches, being more comprehensive and synthesis minded in their approaches:

> "The lawyer must accordingly conceive law not only as an object of cognition and deductive reflection but as an artisan who engages in technical activities, i.e., forging of legal instruments and institutions, and technological activities, i.e., an effort to make the technical instruments more

relevant, more systematic, more rational. Legal science must seek to observe, objectively, the intersubjectivity" of judicial norms instead of being content like classical technological theory, with "prediction" or "certainty."

Quoted from Amselek,
Critiques d'une Reflexion Epistemologique Sur la Theorie duDroit,
Essai de Phenomologia Juridique. 1964.

Medicine

The knowledge base for medicine is primarily the biological sciences and the human sciences. An extensive period of practice and experience is required to enter the profession. Generally, the basic courses are a part of the undergraduate education to prepare for the profession, and practice and clinical experience along with instruction are required as a part of post-graduate education and training. Continuing education is required to maintain competency.

The 19th Century witnessed many changes in medicine as new scientific discoveries, primarily relating to microorganisms, were made. Pasteur (1822-1895) discovered the relationship between bacteria and disease; Joseph Lister (1827-1912) discovered the need for aseptic conditions for surgery; and Robert Koch (1843-1910) discovered the role of toxins in disease. Important developments took place in water supply and waste disposal with the capability to nearly complete eliminate waterborne diseases. Of importance, particular for surgery and dental work was the development of improved anesthetics, particularly ether (1842) and nitrous oxide (1844).

The middle third of the 20th Century was an era of antibiotics, with penicillin (anti-infection) discovered by A. Fleming (1928); streptomycin (antituberculosis) by S. A. Waksman (1944); and poliomyelitis vaccine by Jonas Salk (1954). Research from a wide variety of fields, chemistry, biology, biochemistry, microbiology, physics, and physiology, often done by people who were not physicians, provided the basis for new diagnosis and treatment.

In the 1950s substitute internal organs or transplant of organs, often requiring an artificial or synthetic device, became available. New materials, use of plastics, development of non-invasive instruments required a broader more comprehensive view in the practice of medicine. A metal implant for example to join a broken bone, or as a part of an organ, had to be considered from many aspects to avoid rejection by the body. Organ transplants developed rapidly, with the kidney (1954), lung (1963), liver (1963), pancreas (1966), and heart (1967).

The last quarter of the century saw fantastic developments in instruments for detection and treatment of internal diseases. Many of these developments were hardware-oriented and required the computer to control the device and/ or inform the operator. The CATScan (Computerized Axial Tomography Scanner), by Godfrey Hounsfield (an engineer) and Allen Carmack, for which they received the 1979 Nobel Prize in physiology or medicine, is an example of a device that synthesized many technologies (x-ray, medical, computers) to provide a non-invasive precision medical instrument. This development was followed by the NMR (Nuclear Magnetic Resonance) in the early eighties, the PET (Positron Emission Tomography) scanning, and the MRI (Magnetic Resonance Imaging). The award of a Nobel Prize for the CATScan device signaled a change to considering revolutionary hardware developments for recognition as contrasted to basic science.

Until the latter half of the 20th Century, most physicians practiced as individuals to provide personal health services. There has been a move for a large number of physicians, and other health professionals, to work in clinics or as a member of a group, to provide specialized as well as comprehensive medical services. In the U. S. in 1993, only 37 percent of the physicians were engaged in general practice.

Medical doctors write and talk about the need to consider the entire body, including the psyche, when treating a patient. To carry out this approach is not simple because the approach is related to the education and experience of the doctor. Thus, recently trained versus previously trained doctors, state of biological knowledge, cost and availability of doctors in specialized fields, availability of advanced instruments, and relation to colleagues in other parts of the medical profession are involved. Yet most doctors seem to prescribe "by the book,' often overlooking other aspects of health related to the disease. The medical doctors seem to be deterred from practicing medicine of the "whole" body, because of the threat of malpractice suits that force unneeded tests and may force referrals to specialists. It often appears that doctors are responding more to the legal pressures than to the medical aspects of considering the "whole" body.

Physicians could be classified as primarily analytical in approach, but as the consideration of the whole person has been given more attention, analysis is followed by synthesis. While specialization of physicians has occurred at a rapid pace during the middle of the 20th Century, the need for physicians to consider the system—other organs, other body functions, and non-physiological activities and mental relationships—are in high demand. Specialization has resulted from the high incidence of certain diseases, the production of research results, the large knowledge base and the numerous possibilities or alternatives of treatment.

Veterinarians have many parallels with physicians. Veterinarians often do research with animal analogs to improve human medicine. Some animal parts, such are heart valves, are used for replacement of malfunctioning human valves. In the practice of veterinary medicine, however, the analysis is thwarted (perhaps assisted because verbal misinformation is not given) by a subject unable to discuss the difficulty. In that sense a veterinarian must be quite creative in the synthesis sense to provide appropriate medication or surgery. Another variation is the ethics involved in the practice of the veterinarian as compared to physician, although that difference is being tempered by advocacy groups related to animal care and use.

Other medical fields such as nursing, dentistry, pharmacy, osteopathy, and optometry, have in most respects, changed with the physicians, which include:

» utilizing intensive analysis in identifying diseases or disorders,

» increasing the use of synthesis to improve diagnosis and treatment, not only with respect to biological treatment, but at the same time considering physical, mental and other related impacts, so that a larger system is involved

» being not only concerned with the relationship of the organs of the body, but also the non-physical relationships to health—a more holistic approach,

» integrating mental relationships and social systems to a greater extent than previously,

» helping develop and utilizing specialized equipment to improve analysis of body fluids, to permit non-invasive observations in the body, and to provide carefully controlled treatment. Many of these advances depend on specialized sensors and computers,

» using replacement organs or developing artificial organs to replace diseased organs. This activity requires very extensive knowledge of the human or animal system, much beyond the organ itself, and requires a synthesis of knowledge and experience of several interacting systems acting with the biological system,

» using and developing materials, plastics and alloys and combinations, that are compatible with the biological system, and

» considering alternative methods, such as acupuncture, natural plant and animal chemicals, and

» minimizing the effect of operations on the body

It is a common occurrence to read that the death of a person is caused by a sickness related to a previously diagnosed disease, illness, or operation. Such a statement assumes that studies have been made to determine whether there are connections between the two or more maladies. These statements are usually based on partial or limited knowledge that do not consider the larger system in which the malady or sickness occurs. The interrelationship of two or more possible causes of a malady are available only on a limited basis.

Many of the recent advances in medicine are based on the integration of several disciplines. These advances include replacement of external body parts with manufactured devices (orthopedics), instruments and tools to assist physicians in carrying out surgical techniques (such as minimally invasive surgery for removal of gall bladder or balloon catheter [which is now being restudied] for heart and blood vessels), replacement organs (such as heart and kidney devices), devices to assist the hearing-impaired, seeing-impaired, and mobility-impaired (such devices to replace or assist body functions), and imaging devices such as CATScan and MRI (with heavy involvement of computers) to study vessels, organs and tissues (to provide maps for informing the physicians and for guiding the surgeon). Numerous fields are involved in these developments and challenge the designers. Several of the new developments include miniaturization (particularly for implantable devices and minimally invasive surgery), use of lasers, fiber optics and smaller, safer batteries (batteries evolved from zinc mercuric first used in heart pacemakers, then to lithium, and now to lithium-iodine), and the development of compatible materials (bone splints, skin replacement, body shaping).

Blood substitutes are being sought that will replace human blood, a dream from the 17th Century. (Gillis, 1993) Hopefully, with a substitute blood an adequate safe supply could be maintained for emergency demands resulting from accidents and operations and to meet the needs of those with blood diseases. The substitute blood would not necessarily have the chemical content of human blood, but would serve the function of blood, including excellent oxygen-carrying capacity, good shelf-life, free of immune-triggering agents, and be compatible with all types of blood. (Gillis, 1993) Another approach is to strip off the sugar and protein components from the surface of blood cells to provide a universal blood, type O negative. (Oliwenstein, 1993)

The 20th Century saw a renewed interest by potential medical patients in pursuing a more holistic approach, a greater synthesis, to health. This approach adds to the usual consideration the importance of other factors, such as, e. g., nutrition, environment, and stress. The specialist, reductionist approach with an emphasis on a particular organ, or on the 'disease of the year,' without much reference to the larger system or whole body and its environment is being questioned. The research resulting in the discovery of the structure of DNA and the findings from molecular biology are significantly impacting the practice of medicine.

OTHER APPLIED FIELDS

Agriculture

Harwood (1991) has been one of the recent writers to criticize the research and intellectual activities in agriculture for its almost entire emphasis on the reductionist approach and on specialist work without adequate attention on its impact on the stifling of synthesis. Each agricultural enterprise has been treated as if it functioned by itself, without reference to biological interactions of different crops, without reference to the impact of the crop or its system on the larger systems of energy and environment, almost entirely on the production aspects of the crop. A few people recognize this limitation, and programs such as integrated pest management have developed to look at the larger system. Harwood (1991) claims that the research as being devoted to more and more descriptive aspects of ever smaller and more basic pieces of organisms and units of the system. The development of technology becomes more and more a matter of small incremental changes in existing materials and processes. With such thinking there is a decreased capacity for one to think outside the conventional mainstream. He recommends that "a process for creative synthesis must be added to the growing excellence in specialist focus." He sees that:

» present procedures of specialists are stifling synthesis and systems study

» the reductionists protocol is a club wielded over systems efforts

» scientific careers are overly dominated by the need to publish through the narrow channels based on academic disciplines

» there is well-grounded research on integrated systems for which methodologies need to be developed

The observations relating to agriculture apply to other fields as well. Along with other applied fields development of methodologies remains a major challenge to those who want to use analysis as part of a larger system to invoke synthesis as an approach.

Business

A person with a Ph. D. in management explained that the most difficult question asked before his examining committee was, "What is business?" Most of us believe that we know what business is, but find it difficult to put into words. Generally, business can be described as a purposeful activity, the activity directed to some end, such as making a profit, as for a commercial or industrial enterprise. In terms of a commercial enterprise, business includes economics and finance, organization and management. Both of these include many parts. A major aspect in business is to synthesize many parts, but also to fit into a larger synthesis.

Many changes are occurring in business that increase the need for additional emphasis on synthesis in the future:

» Increase in the size of companies, particularly combining already large companies,

» Increase in size and location of geographical areas, from local to regional, to national, to global activities, considering not only the source of raw materials but also markets,

» Increase in multi-national companies, usually including above aspects,

» Increase in public involvement of private companies, particularly as relates to working conditions, environmental impact, consumer (not limited to the customer) activism, services for non-employees, and as a tool to carry out government programs and policies, and

» Increase in number and level of skills needed to operate the company.

Managing companies has become more challenging than ever before as companies have become responsible and affected by laws such as those that specify internal procedures, hiring policies, affirmative action, insurance and pension requirements, and external procedures such as trade with foreign countries, or prohibition of trade or monetary policies to force other countries to take certain actions, usually related to freedom of the population. These activities force a larger synthesis on management than in the past. To the 'bean counter' mentality must be added synthesis as a way of thinking and a way of acting.

Two segments of business involve professions as identified by education, examination, experience, and licensing by government, paralleling registration for the public practice of engineering, P. E. These are the Certified Public Accountant (CPA) for accounting, primarily analysis, and the lawyer for legal matters. Most of the remainder of business activities involve the application of "art and science," with more emphasis on practice than on theory. The influence of uncontrollable, often unpredictable (yet knowable possibilities) events challenge those in the field to continually analyze, evaluate, and synthesize the effect of various combinations of factors on business. Events such as war, disasters (local and global), assassination, terrorism, previously unknown diseases (such as AIDS) demand that predominately analytic approaches be combined with synthesis to prepare a business for the future. Few fields of activities have greater justification than business for a synthesis approach.

Business practice 'aimed toward action and the pursuit of practical aims' such as making a profit and business philosophy for theory establishes the framework for looking at business in terms of analysis and synthesis.' (Lodge, 1945) The first part of business is sociology considered as a science, and the second includes primarily realism, idealism, and pragmatism. (Image 6-2). "Realism holds that physical things exist in their own right, prior to and independently of our awareness of them. Idealism is based on the premise that the mind is central and materials are secondary. Pragmatism is based on the priority of practical reason, including empiricism, instrumentalism, and experimentalism." Combining the appropriate elements of the above into a single body of doctrine... integrating the insights of distinctive philosophies... emasculates each of the original vital points of view... the synthesis method is hardly the best conceivable way to realize this ideal." (Lodge, 1945) But combined with a strong analysis, incorporating synthesis into an analytic-synthesis method can be a productive useful approach not only for business but for other applied areas as well.

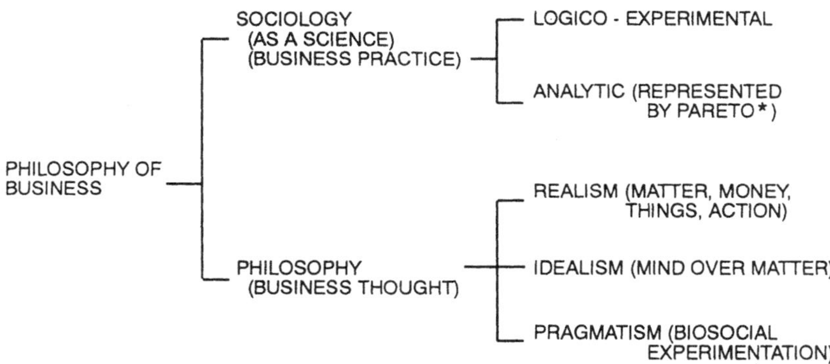

Image 6-2. Philosophy of Business (Lodge, 1945)

In business, the practice gets the attention: the CEO with high income, the person who makes a 'killing' in the stock market, and the investor who makes a high return on investment. Many of these actions are based on short term objectives, based on a quarterly or an annual profit line, where analysis is a very effective tool. The individual incentives for long term goals are minimal. Until a larger number of people—politicians, investors, managers—get involved in a greater 'synthesis' approach, decisions for long term economic health will come out second best. Bonuses should be given after a 3-5 year period so that long term management decisions are encouraged.

In a remarkable book, Senge (1990) writes of systems thinking as the fifth discipline. This is a straight forward support of synthesis as a way of thinking and a way of doing. Systems thinking (or synthesis) is an alternative to the reductionism so prevalent in today's business. The systems approach is one of openness, breaking down the walls of disciplines, emphasizes process thinking (rather than product), and opens up consideration of effects of Eastern and Western culture, and integrates reason and intuition. Systems thinking is explained further as being a pyramid made up of, at the bottom, practices (system archetypes, simulation); in the middle, principles (structure influences

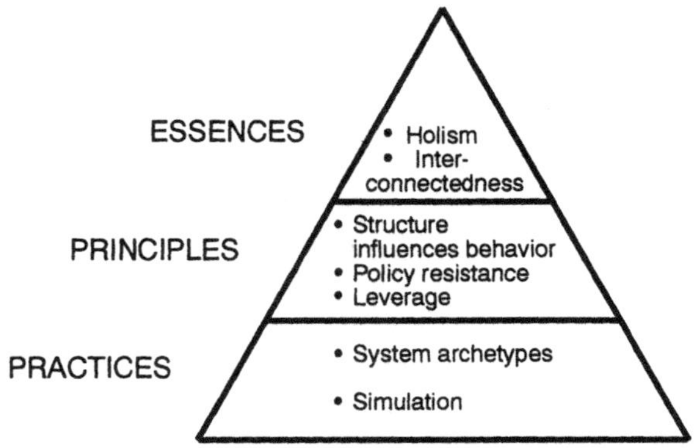

Image 6-3. Systems Thinking (Senge, 1990)

behavior, policy resistance, leverage); and convergence at the top into essences (holism, interconnectedness). (Senge, 1990) (Image 6-3) This approach, expressed in different terminology represents the combined analytic-synthetic concept, the needed and emerging approach in business.

Leaders are beginning to recognize to a greater extent than ever before the value of this new systems or synthesis approach. Books, short courses, workshops, seminars, and discussions on the subject are more commonplace than ever before and promises to continue to increase in importance and popularity. A report by the American Council on Education (1985) admonishes business schools "to do a better job of addressing the complex interrelatedness of most issues facing business," to address the "resolution of interdisciplinary concerns," and for faculty and students to "understand the total business process as well as the role their particular specialty plays in that process," a clarion call for strengthening the analysis-synthesis approach.

Leadership will act more responsibly and productively and those in the organization will heed the call and favorably influence the output-production (goods and services) line. Both leaders and followers, executives and managers, will incorporate more factors, more variables in a responsible manner, in the thinking and the doing, as a joint effort. Management is particularly effective working within a paradigm (shifts of thinking for self and organization), and doing; executives should be particularly effective in working between paradigms. (Image 6-4).

Education

Synthesis as an approach has much to offer education, particularly as complementary to analysis. Beginning learning starts with a "holistic" view, an in-

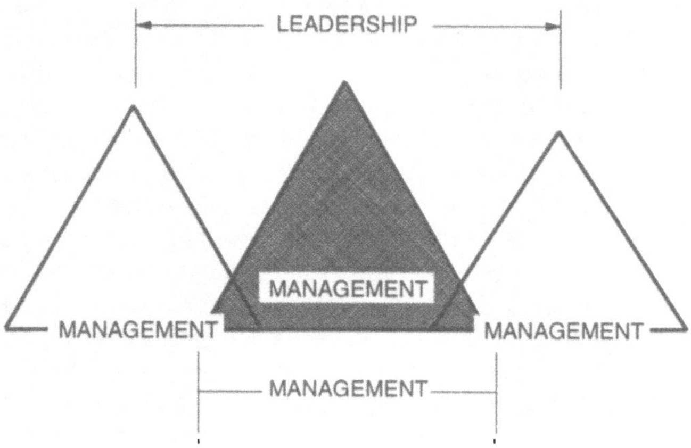

Image 6-4. Paradigms of Leadership and Management

born synthesis of things. In natural things, branches, trunks, and leaves are seen as trees. Legs, body, and head are seen as animals or people. With artificial things, whether artwork, vehicles, structures, a child sees a painting, fire engine or bridge. I knew many varieties of trees by seeing them in the forest, a whole view. It was not until I took a course in botany that I could more specifically identify certain trees or varieties of trees. Then I was taught to observe details of such things as serrations of leaves, number and location of leaves on the stems and branches, shape and makeup of blossoms, and color or shade or smoothness of the bark and branches. A general observation of the whole was enhanced by specialized view of the observed. Serious students and researchers become deeply engrossed in studying the details, analysis of the idea or thing, and often continue their studies for life in that vein. The detailed analytic studies could be enhanced to serve a wider audience by incorporating to a greater extent, a synthesis approach.

Without detailed analysis, synthesis would be shallow. A greater use and value of analysis would occur with increased use of synthesis. However, the value of synthesis as a way of thinking, as a supplement to analysis, needs strengthening and should be inculcated into teaching and learning. Strengthening synthesis as a part of education should not be an excuse for decreasing analysis or for using false analytical or factual data. The recent revelation of numerous high school textbooks with incorrect factual data is unacceptable for both analytical and synthetical approaches. The thousands of errors in fact cannot be dismissed on the basis that facts aren't important in history, that the big picture is most important. But that big picture must include correct available facts whether in the natural or unnatural or artificial world. Being politically correct (PC) without consideration of the facts of the situation is not acceptable.

In teaching people to read, controversy exists on whether to teach the student to read by whole words, called 'sight reading' or 'whole language' or 'look-say', rather than to teach the alphabet, the A, B, Cs. (Bower, 1992) The first might be considered a synthesis approach and the second an analysis approach. A combination of the two is required, and whether analytic-synthetic or synthetic-analytic is most effective probably depends largely on the student, but both modes are needed, an example of spiraling. A similar example could be given involving numbers and learning to divide, multiply, add, and subtract.

Teachers build their instruction on the innate curiosities of students, taking the capability of seeing the whole and then challenging students to utilize their curiosity to develop analytical skills. Developing those analytical skills continues on through secondary schools and college. Certainly in college, if not before, an emphasis on synthesis needs to be reintroduced to prepare the student for the challenges to face them as there will be a shift from individual to team efforts, from disciplinary to multidisciplinary efforts, increased interaction between natural and artificial systems, and involvement in larger systems. More and more, increased knowledge and understanding will result from increasing the spiraling of interaction of analysis and synthesis. Bruner (1963) refers to the value of a 'spiral curriculum' that turns back on itself at higher levels. That approach would also help satisfy the need for analysis and synthesis to be appropriately considered in the learning process. The failure to consider these approaches leads to criticisms of education without relevance and the lack of involvement of many students to prepare for future service to society. The realist teacher tends to use the synthetic method in presenting material by showing how the whole emerges using the elements and details.

The success of the teacher often depends on understanding the characteristics of the students. Lodge (1947) describes three major student categories: pragmatist, idealist, and realist. The pragmatist student has practical interests, sees life as one thing after another, not a vaguely conceived whole, and looks on tomorrow as a new day. This student concentrates on what is now happening then moves to something else. This student is action oriented, is always doing something, is interested in surrounding life and is not aloof. The idealist student is vague, shy, quiet, with an uncertainty about the future, perhaps dreamy. This student is usually conscious of an inner self, but has difficulty envisioning its nature. This student looks for goodness and beauty and truth in people and is very sensitive to personal relations, is strongly influenced by others, particularly by teachers. The realist student approaches learning with definite ideas and demands. This student expects teachers to know and how to present their subjects. Thes student expects classes to cover the subjects for which the course is designed, with little peripheral coverage, not so much interested in art, history, or languages unless these relate to the field of study.

Major universities are attempting to become more effective and efficient financially by restructuring the organization, reducing administrative and service costs, moving away from an overemphasis on research, and in some cases combining appropriate units to provide multidisciplinary units. All of these require a combination of analysis and synthesis approaches, the latter of which is not the usual skill of most people in academia.

Small colleges in particular, usually private, are facing difficult times financially. These colleges often try to duplicate the offerings of the larger, often prestigious colleges and universities by incorporating a large number of courses, usually analytical in nature, and insisting on a heavy faculty commitment to research. Research is accepted as a way of keeping current in a field while contributing to the literature. But there are alternatives, some of which could be more appropriate for small colleges. Limiting the number of programs, limiting the range of activities that often include international activities not aimed at student learning, limiting the laboratory research and limiting off-campus centers are approaches. A philosophical approach is to emphasize a synthesis approach—thus bringing in the appropriate aspects of all possible activities, while decreasing the analytic approach that is being done so well by others, usually individuals. Academic respectability could be gained by writing comprehensive review papers on appropriate subjects, a practice that would improve the breadth of the knowledge base for the teacher; organizing and participating in interdisciplinary meetings, that would provide a better context for the student upon graduation; and be a leader for future research by others, by identifying gaps in knowledge. Thus, a unique education would be provided and the students would be prepared to carry a mantle of leadership and provide an important human resource for the future.

The Bloom's taxonomy and the Newcomb-Trefz Model represent two approaches to domains of learning. Both of these models include synthesis as an important aspect of learning, building upon analysis. The Bloom's (Bloom, 1956) taxonomy includes the following six steps: A. Knowledge, B. Comprehension, C. Application, D. Analysis, E. Synthesis, and F. Evaluation. Evaluation is presented as quantitative and qualitative judgments about the extent to which the material and methods produced satisfy the criteria. The judgments are made in terms of internal evidence and in terms of external criteria. The latter is established on the basis of "user" (students) needs as well.

The Newcomb-Trefz Model includes all of the items of the Bloom's taxonomy but in four steps or levels: I. Remembering, II. Processing, III. Creating, and IV. Evaluating. (Newcomb and Trefz, 1987) Creating is considered to include or be part of synthesis. (Image 6-5)

The cognitive (perception, awareness, associated) (Matrix 6-1) levels are:

1. Remembering — involves a recall of the information presented to the student,

2. Processing — requires an understanding of the information and its use,

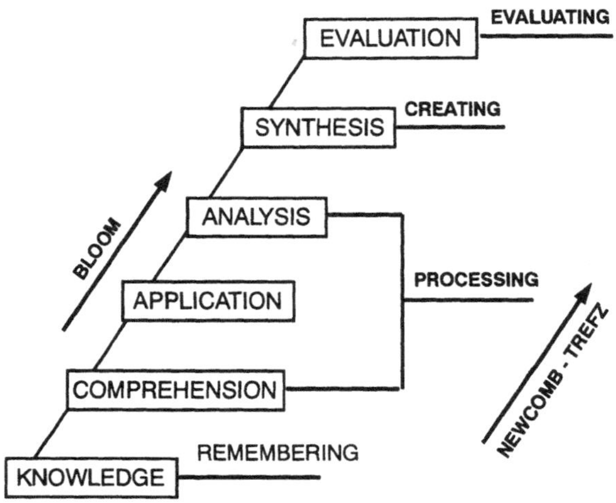

Image 6-5. Newcomb-Trefz Comparison to Bloom's Taxonomy

3. Creating — requires the combination of information in a form new to the student, and
4. Evaluating — requires the student to use the information in making an independent judgment.

Many messages are conveyed by images, in education, politics, media and business, as well as by the artifacts that we produce. One image (picture) can be accepted as representative of the whole, when in fact it may represent the opposite of the truth or accurate representation. The opinions of millions of people can be falsely founded by the constant bombardment of an unrepresentative image of what is, with a distorted view of the world or parts of it. It is impossible for everybody to get all the information on a subject, but we should seek a larger view than is normally handed to us. With instant communication on a subject, people can be overwhelmed by incorrect information, and be pressured to express opinions or judgments without reasoned consideration. The education system should equip people to seek a larger, more complete view, synthesizing as well as analyzing information and knowledge.

Social Sciences

The social and behavioral sciences are anthropology, psychology, sociology, political science and economics. Some observations by Rigney and Barnes (1980) provide a view of some of these subjects. Although sociology, in principle, has been a 'synthetic discipline, more often in practice it has served as an institial discipline (filling gaps).' They state that 'anthropology as a holistic discipline has been among the most interdisciplinary in the social sciences.'

Economics has been relatively isolated from its neighboring disciplines. Economics and psychology, two disciplines with the largest literature also have the highest self-citation, with 80 percent of the references from their own fields. Sociology, anthropology, and political sciences have a self-citation of about 50 percent.

Matrix 6-1. Vocabulary Representing Various Cognitive Levels [See Image 6-5]

Cognitive Level				
Remembering	Processing		Creating	Evaluating
acquire	analyze	infer	combine	appraise
cite	apply	interpolate	compose	argue
define	associate	interpret	constitut	assess
identify	categorize	outline	construct	conclude
label	change	paraphrase	create	consider
list	choose	point out	derive	decide
name	classify	predict	design	evaluate
recall	compare	prepare	develop	judge
recite	compute	rearrange	devise	standardize
recognize	contrast	relate	document	validate
reproduce	convert	reorder	formulate	weigh
state	deduce	rephrase	integrate	
	demonstrate	represent	modify	
	describe	restate	originate	
	detect	restructure	organize	
	determine	summarize	plan	
	differentiate	transfer	produce	
	discriminate	transform	propose	
	distinguish	translate	reorganize	
	draw	use	revise	
	estimate		rewrite	
	explain		specify	
	extend		synthesize	
	extrapolate		tell	
	generalize		transmit	
	illustrate		write	

Source: Newcomb and Trefz. 1987.

Adapted from Chamberlain, V. M. and Kelly, J. M. 1981. Creative Home Economics Instruction. McGraw-Hill Book Co., New York, NY; Clegg, A. A. 1968. Teach strategies of questioning for eliciting selected and cognitive student response. Applied Research Training Program, Research Report No. 1. University of Massachusetts, Amherst, MA; Hall, K. S. 1983. Content structuring and question-asking for computer-based instructional objectives and classification schemes. Electrical Technology 24(7):21-26.

When one attempts to synthesize a bigger picture, a larger system, beyond those in which firm analytical data are available, as for example in political systems, there is introduced the possibility of making wrong or incorrect judgments. There is a tendency to be politically correct (PC), that is to incorporate the latest politically-based opinions. Thus, analytical data and information can be overridden, denigrated, or overshadowed by others. When considering a bigger system one should move a step at a time, like peeling an onion, incorporating analytic data available at each level, and integrating with previous layers in the synthesis. These steps of assembling a system would normally follow the more visionary approach of having an end product or process in mind. In synthesis there are usually more variables, more unknowns, less controllables, thus presenting more difficulties, taking more time and providing more risks, while making possible creative and innovative results.

An interesting example of the relationship of parts to the whole is written by Catherine Drinker Bowen (1966). The leaders of the Revolutionary War undoubtedly had a hazy vision of the structure of government they wanted to provide the citizens for the freedom they sought. The Confederation of States lacked the needed cohesiveness and strength of government. The importance of appropriate connections between the parts (states) and the whole (nation) is stated by James Wilson in 1787 at the Constitutional Convention:

> "No sooner were the state governments formed than their jealousy and ambition began to display themselves. Each endeavored to cut a slice from the common loaf to add to its own morsel, till at length the Confederation became frittered down to the impotent condition in which it now stands. Review the progress of the Articles of Confederation through Congress and compare the first and last draft of it! ...One of its vices is the want of an effectual control in the whole over its parts. What danger is there that the whole will unnecessarily sacrifice a part? But reverse the case, and leave the whole at the mercy of each part, and will not the general interest be continually sacrificed to local interests?"

Cognitive studies in psychology is called knowledge integration rather than synthesis. People working in cognitive skills have taken the approach of designing experiments to exclude any kind of integration of knowledge from previous experience. In research, if one were testing for text comprehension or recall, previous knowledge is excluded, so a priori is avoided. Hume believed that scientific studies in psychology were not 'experimental' but were 'experiential.'

The computer helps do on-line cognitive processes, to manipulate and explore knowledge bases, and in this sense, cognitive studies have focused on analytical thinking. Early psychology had an incomplete view, so natural psychology (in everyday use) has given way to contextural psychology, which has a place for synthetic approaches, after 150 years of emphasizing analytic approaches.

Recent worldwide recognition for economists has increasingly gone to people looking at world models rather than local models, particularly for Nobel prize laureates. In 1993 two people (R. W. Fogel and D. C. North) were recognized for their efforts to combine the often abstract methods of economists with the historian's attention to empirical detail. By recognizing these people, some interpreted the selection as representative of pragmatism in economics, and a move away from the statistical and mathematical theory and modeling. Of further interest, however, is that the approaches of the two people are quite different. The first finds a fact, proves that it is a fact, and then proves it is a fact for numerous applications. The second discovers a theory, then fills it in, to show how the theory helps to understand an event.

Economists and engineers should be close allies. A former graduate student of mine became the Minister of Treasury in his country. I remarked to him how unusual it was for an engineer to reach this position. Some politics was probably involved, but he told me that the job was challenging and that he felt comfortable in the work. He said the job required using some of the principles that I had taught in heat transfer, Btus in and Btus out as compared to money in and money out. Most engineering projects in a capitalistic society are based on economics, and a powerful tool is the input-output, or for some applications input-outcome (I-O) model. Although, as the name applies, it is a simple model, but doesn't cover many aspects that society wants accomplished where it is difficult to quantify variables, such as a clean environment, reduction of dependence on fossil fuels, and safety in the workplace. The model does not work too well in resource allocation. A major effort is involved in determining the cost involved by internalizing externalities. Modification of the I-O model to represent in monetary terms the costs and benefits of these aspects becomes the basis for considerable discussion. Technologies in which a new process or a replacement process or product are compared to determine the potential profitability, the impact on the environment, and the ability to meet the demand of the user.

SUMMARY

The professions, particularly the learned professions, by reason of their responsibility to serve the public, have a major responsibility to synthesize information to help improve existing systems, natural or artificial. The professions should be looked upon as human endeavors as well as functioning to design systems that consider the human and social environment. Educational activities need to be developed to improve the synthesis activities of those who practice in these fields. Those in the professions can use synthesis to identify gaps in the knowledge base. Those in the arts and sciences can help supply information to meet the needs of a system larger than they would ordinarily consider.

Architecture has always been associated with the arts and the education of architects emphasizes that aspect and gives a more holistic experience than most engineering education. Challengers to conventional engineering education press for greater attention to synthesis, building on a strong analysis. The synthesis is represented by the integration of subjects, technical and non-technical, without necessarily homogenation of topics. The approach presents a real challenge in that most educational institutions want to accomplish these broader objectives without adding more time for students to complete their studies.

The legal profession is based on many activities that would be identified as synthesis, both in principles and practices. Analytical jurisprudence represents the study of law in terms of its logical structure. Although the term synthetic jurisprudence is not used in the law literature, synthesis is involved in practically all aspects, that is relating to a larger body of knowledge, and developing presentations that meet the need of the client or customer.

Education in the medical profession has emphasized clinical work, an approach from which engineers could benefit, and only in recent years has built up the basic biochemical sciences, including nutrition, into its programs. Many advances are occurring in medicine based on new methods, tools, and biochemical knowledge in which a greater synthesis of subjects involving many other fields, particularly engineering, will challenge the practitioner and benefit the user. The client will particularly benefit by avoiding surgery, less invasive surgery, non-invasive diagnostic equipment, monitoring devices, and replacement of body parts. More effort should be devoted to educating students at all levels regarding understanding the importance of incorporating synthesis. Images are powerful tools for developing thinking and visualizing the possible. However, if the image does not properly represent the part in terms of the larger system, a distorted or incorrect situation is presented and often believed. Thus, a picture of an isolated event, without presenting the larger view, can be misleading. Advertising, media presentations, sales pitches, and political messages are subject to such practices. A conscious effort to present and incorporate synthesis and analysis on a balanced basis is necessary, not only for technological but sociological progress.

Part 7.
Synthesis of Artificial Systems (Engineering), Philosophy and Application

Increasingly, artificial systems are becoming a part of our technological society and will increase in importance and utilization in the coming century. Synthesis is a major activity in the creation of unnatural or artificial components and systems and is a way of thinking as well as doing or making. This section will emphasize artificial systems in contrast to artwork and literature, covered previously. Philosophical concepts, such as emergent properties and teleological approaches, are used for describing artificial systems that were not in mind at the time the concepts were developed. Although the word 'systems' is strongly ingrained in our discussions there is not a clear description of a 'system' in the generic sense that applies to the numerous uses. An artifact, a made or crafted thing, such as a painting, an alloy, a gear, or imitated items, could be considered by some people as a system. However, in the usual usage of things those items would not be considered as systems. One might think of a system or systems within systems, Image 7-1. Thus, there are many systems within an airplane, train, ship, truck and automobile, yet all of these contribute to or are a part of a transportation system. One might question as to whether a pavement is a system, but certainly the pavement is an important component in a transportation system. In earlier parlance we referred to components within sub-systems within systems. Although still applicable, we can enlarge our consideration to larger and larger systems, and more and more components and sub-systems. As we expand our thinking to consider a larger system, systems become sub-systems.

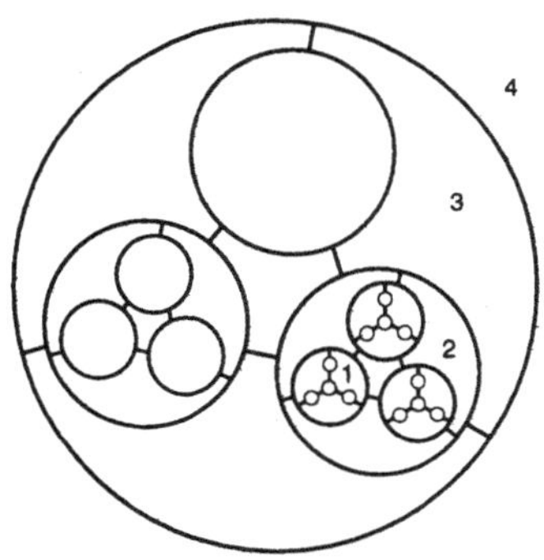

Image 7-1. Systems Within a System

EMERGENT AND ANIMISM

The concepts of emergent and animism were developed with natural systems in mind, but many aspects can be applied to unnatural or artificial systems. Angeles (1981) defines an emergent property as:

> "The new qualitative synthesis produced by structures organized in certain patterns that cannot be predicted from the examination of the constituent parts of the whole. . .The emergent quality of water could not have been predicted from the knowledge that two parts of hydrogen and one part of oxygen, both combustible gases, would produce a liquid."

NATURAL LAWS

Natural laws have been developed to describe natural substances and systems. These laws evolve to include more and more variables as research is done and more is learned. In the case of water, for example, more and more is learned about the elements and their interaction so more and more predictions can be made about the properties of the combined elements to make water. Many of the natural laws are useful in predicting the response of synthetic materials and artificial systems but the known physical laws often fall short in describing an artificial system. Elements, compounds, and materials are assembled in new ways and until experiments are conducted, the emergent properties of the artificial system do not permit one to know with much certainty the response of the system. Engineering research is often involved in experimenting with unnatural or artificial systems to determine their properties.

The flow of electrons in a transistor was impossible to predict even though much research had been done on electrons. Experimentation provided modifications to existing physical laws or resulted in the development of new laws so that predictive models could be developed. The flow of electrons through memory chips could not be predicted with the knowledge available at the time. The flow of electrons in nanometer size of circuits was different from the flow of electricity in power circuits so that research was needed to develop the relationships.

The flow of air over large or creatively shaped wings of an airplane could be predicted but not verified without considerable experimentation. Whereas a scientist might do research in these areas because of curiosity, and call the work 'basic,' the same work done by an engineer to meet objectives of fulfilling the synthesis of a part of the system, might be called 'applied.' The vast amount of literature attempting to elevate basic research above applied research is often misleading and provides a disservice to the public enterprise. Research should be considered more in a synthesis mode, utilizing people working in partnerships to identify and solve problems, involving the potential user, client, and recipient of the results.

RESPONSE OF ARTIFICIAL/SYNTHETIC

A bridge is made of materials and components for which the basic physical properties are reasonably well known. However, one isn't secure or comfortable in building a large bridge based only on the knowledge of component properties. The interactions between and among components can be more complicated than the components themselves, requiring research, development, and field tests to evaluate. As new materials are considered, experimentation is required to evaluate the response of the system. The size of the bridge is a critical parameter and we are not secure in the design if we increase the size by more than one or two orders of magnitude without additional testing because different interactions often occur. Unanticipated external factors can come into play that could cause failure.

Standardization of design (after some experience) provides an opportunity for data and information to be collected that can help improve the safety and decrease the cost of the structure. Standardization often leads to "handbook" data that tends to decrease creativity for future designs. Many large structures were built with a paucity of data which meant that the structure was over-built or the safety would be compromised, and the cost much more than would have been necessary or the structure was often under built and failed. Large chemical processing plants, complicated biotechnological operations, and comprehensive nuclear power plants are but a few of the examples of large, complicated artifacts that can not be built based on scientific laws by themselves. The task of the researcher dealing with artificial systems is quite complicated because of the many variables, and the possibility of imposing additional variables during the process of creative design.

NUCLEAR POWER REACTOR

One of the reasons that the nuclear powered submarine project was successful is that the design was fixed after an appropriate experimental period, the length of which is based on subjective information, experience, and judgment. Admiral H. G. Rickover (1900-1986) is usually credited with managing the development of the Navy submarine, U. S. S. Nautilus, launched in 1954. In contrast, nuclear power systems for electric power production, even though of the same type, i.e. pressurized water reactors (PWR), were designed and built with different specifications and of various sizes for different power plants, such that the control, maintenance, operation and safety were not fixed. Thus, complete system tests, operator training, performance evaluation, and safety procedures had to be entirely redone for each new reactor, adding considerable cost and additional time to get each new unit into operation. In many situations neither the budget nor time to get into operation were met, two important criteria for engineering projects. The performance of that segment of our industry is not as efficient and acceptable as desired, but is not entirely due to scientific or technological reasons. The concept of fail-safe reactors, if pursued and supported earlier, could have led to better acceptance, both technologically and politically, and established a more confident public regarding use of nuclear power for production of electricity.

LARGE ARTIFICIAL SYSTEMS

Many of the large structures designed and built by engineers are studied in much the same way that oceanologists and meteorologists learn by "observational" science. They are not in a position to move a planet to determine the effect on gravity or tides, so must observe the happenings and attempt to quantify and explain results. Similar relationships exist for the study of earthquakes and volcanoes, weather, and large water bodies, often involving engineers.

Future works of engineers will incorporate sensors in structures so that appropriate data can be obtained from the structure during its use, in different environments, and under various loadings, mimicking observational science. An intensively instrumented structure or system can provide considerable data to monitor the "life" of the system from which the "health" of the structure can be assured, maintenance prescribed, failure mitigated and life extended. These data can then lead one step closer to teleological animism, defined by Angeles (1981):

> " The theory that (a) parts are adapted (coordinated, adjusted) to other parts within a whole in order to achieve the purpose of the whole, and (b) the whole itself sets the purpose and causes the adaption."
>
> " Teleology is the study of phenomena exhibiting order, design, purposes, ends, goals, tendencies, aims, direction, and how they are achieved in the process of development."

The concepts of emergent and animism were developed for natural systems, but can be projected to be applicable to unnatural or artificial systems.

VISUALIZATION

Visualization means forming a mental image of some thought or idea generally not being seen. Words can be the source of images that get their meaning from thoughts or ideas, or vice-versa. It was Aristotle who is believed to have said that spoken words are the symbols of mental experience and written words are the symbols of spoken words. This ties into the theme of Thoughts (ideas) to Theses (words—spoken or written) to Things (products). Visualization—forming mental images—provides a means of carrying imagination one step closer to reality; it greatly enlarges human experience, and takes us from a narrow present to an enlarged, perhaps infinite future.

Visualization has played a major role in science, engineering, architecture, technology, and art. The success of people in these fields depends on their ability to visualize, whether theory, process, or product, with the visualization gradually changing as the image moves from concept to product—like a computer melding of various inputs to come up with a computer simulation of a person's picture.

Examples in the scientific world are Bohr's visualization of considering an atom as a miniature Copernican system, although inaccurate, led to many advances; Fermi's visualization of how the neutron decayed to a proton; and Einstein's visualization of relativity based on appearance of motion of trains approaching or separating from each other based on the position of the observer. In the technological world, whether it's making a horseshoe, an appliance for shop or kitchen, an instrument for writing, or clothes, or designing a toy or a game, as examples, there is a visualization that leads to the product.

Visualization helps evaluate the perspective of the object to the viewer, the fit of parts, the interference of parts, to monitor a mathematical function during calculations, and to get a 'feeling' as to order of magnitude. Engineers who learned with the slide rule at their side, now antiquated, had a better perception of order of magnitude than many who followed, and were more likely to avoid unreasonable number values in calculations.

Visualization has taken on a new meaning with the availability and use of the computer. The computer system provides a large memory of data and information, rapid response for handling massive data, ability to compute rapidly, and various means of displaying graphically or pictorially the results. Understanding and insight are obtained using computer based approaches. Some laboratory experimentation can be replaced by computer simulation. Simulation can be used not only for physical phenomena but for relationship that can be imitated or imagined, thus producing a visual result. Although the creative effort is contained in an individual's mental capability, the ability to analyze and synthesize is greatly enhanced by the computer. The visualization of scientific data is done with the objective of understanding the relationships, as contrasted to graphics that are used to represent results already understood. Computer software permits one to monitor visually (during the process) the solution of a mathematically expressed problem. One can 'play around' with the variables to find different or more effective results. Interdisciplinary and multidisciplinary teams are involved in developing the hardware and software, in bringing various expertise together for defining and solving problems involving large data sets. Interestingly, one of the first major commercial used of the supercomputers was for making (synthesizing) pictures and films for the television and movie industry. Large artifacts such as buildings, airplanes, automobiles, and particularly large systems such as transportation, communications, and energy, utilize visualization techniques based on computer technology. These larger systems involve even more analysis than previously, but must involve considerably more synthesis, particularly as a way of thinking, than ever before.

SYNTHESIS AS NATURAL AND ARTIFICIAL PROCESSES

Synthesis of things, individual items and systems, can be a natural process, such as in photosynthesis, and can be an unnatural or artificial process, as in the design and making of a device, vehicle, or building. For natural systems, the driving force is the energy on the input side and the environment on the output side. For artificial systems, the driving force is motivated by the end uses that the designer and manufacturer have in mind.

Through natural processes biological components and systems are produced. Initially, one might think that it would be more difficult to distinguish the difference between components and parts in a biological system. But that is not so with higher life systems such as plants, animals, and people. In plants the roots, trunks, branches, stems and leaves are rather distinct subsystems of a larger system. For animals and people, toes, feet, legs, body, neck, and head are usually well-identified components. In their own way, new species are evolved which replicate themselves, usually more accurately than artificial systems. Through self-assembly, with appropriate components in an appropriate environment, compounds and more complicated 'higher' inanimate materials are formed. Although humans can interfere in this process, the "rules of nature" control the events to synthesize a product. Often we have to discover the rules of nature in order to proceed.

It is for the inanimate objects made by human beings that new systems are synthesized and duplicated by building, constructing, and manufacturing. Here imagination is fulfilled with synthesis of an infinite variety of items-from very small to very large systems. The rapidly developing field of nanotechnology, the small (measured in microns, in which one micron is 10^{-6} meter or Angstroms, in which one Angstrom is 10^{-10} meter) to the very large. Every year brings new, larger, more comprehensive machine, structure, and communication systems. All of these systems, whether small or large, involve complexity, interconnectivity, interfacing and integration that are the handmaidens of synthesis. The rules of synthesis need to be developed, which as a starting point are the rules of integration and systems.

INTENTIONALITY

Synthesis is a result of intentionality (a part of phenomenality; relates to design), the ability of consciousness to create a mental object that need not exist in the physical world, apply its content to reality, and direct activity toward results. This compares to the THOUGHTS – to – THESES – to – THINGS concept presented previously. In the natural domain, teleological animism is the theory that parts are adapted to other parts within a whole in order to achieve the purpose of the whole and the whole itself sets the purpose and causes the adaptation. This relates to teleology, the study of phenomena exhibiting order, design, purposes, ends, goals, tendencies, aims, directions, and how they are achieved in a process of development. (Angeles, 1981)

NEMAWASHI — A SYNTHESIS (JAPANESE)

The observation of the Japanese procedure in which a group discussion often occurs before an answer is given to a question leads one to think that the Japanese approach is fundamentally more synthetic than analytic. As discussed with Dr. Kasuke Takahashi, the process is called Nemawashi, which is "root" and "rotate around." (Image 7-2) Thus, the discussion comes from each person

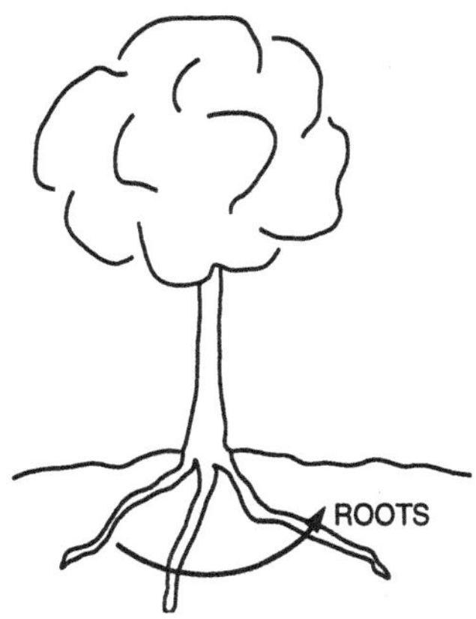

Image 7-2. Nemawashi

as the root of an idea, and the identity of the original ideas developed become a part of the whole, a synthesis of the thoughts of many, and is applied to political and scientific thinking, a part of the Japanese culture. A longer time is usually required to come to a decision. Once the decision is made all work together to carry out the task because it is not based on a limited majority opinion, but with all people agreeing. This group activity, like engineering, is not the way people are generally recognized for Nobel prizes.

DEMINĠ — FROM ANALYST TO ANALYST-SYNTHESIST

The approach by Dr. W. Edwards Deming (1900-1993) was originally analytic and filled a major need in Japanese manufacturing. With time Dr. Deming's approach moved more to synthesis, and today, with his emphasis on quality he would be considered a synthesist. Analysis has dominated U. S. manufacturing where it served well for many years, but the critical need for the future is for increased attention to synthesis while maintaining the strength in analysis. Dr. Peter F. Drucker, basically a synthesizer rather than an analyzer, provided that approach to U. S. business management, but the technological counterpart for manufacturing (synthesis) was lacking in the U. S. until Dr. Deming's contributions became recognized and many others have joined in the efforts to improve synthesis.

Another aspect of the Japanese is that there is a lack of precision in their language and culture, even less than with the Chinese, which is manifested in their approach to manufacturing. Western methods, first the Germans and then other Western countries, provided guidance for developing precision in technology, as evidenced by formation of Precision Engineering faculties in Japan in the late thirties by German professors. Generally, precision is associated with analysis (measurement to 0.0001, 0.0002, etc.). The lack of precision could imply a stronger synthesis (order of magnitude measurement or representation, such as 10^0, 10^1, 10^2, etc.). Both approaches are required, depending on the circumstances, and a balanced approach is necessary. Care must be taken to avoid confusing precision with accuracy.

PRECISION

The vernier scale provides a method of designing precision measuring instruments by providing greater accuracy in reading the measurements. The vernier scale and associated instruments for measuring lengths and angles were named after their inventor, Pierre Vernier (1580-1637), a French mathematician. The scale and the associated instruments using the scale are still in use and one wonders whether the principle would have been discovered if Vernier had not developed it in the early 1600s, nearly 400 years ago. The vernier has a short graduated scale that slides beside the larger scale, with ten divisions on the small scale equal to nine divisions on the large scale. To use the vernier, the large scale (ruler, calipers, micrometer) is used to measure the part. The small scale is then laid out to the end of the measurement and the line of the small scale that somes closest to lining up with the large scale gives the reading with an additional digit of accuracy. Thus, a reading of 30 millimeters on the large scale, plus a reading of four (at the matching lines) on the small scale gives a reading of 30.4. There are modifications (different ratios of small to large scale), accessories (magnifying glasses), and adaption directly to the tool (for control as well as measuring). The principle of the vernier scale was taught in engineering classes when I was an undergraduate engineering student fifty years ago. I wonder how many engineers could identify, much less explain the vernier scale today?

SYSTEMIZATION

Systemization is a representation of synthesis; systemization thinking has increased scope of thinking to the holistic. Integration is a step towards synthesis. Interfacing, interdependence, interdisciplinary (multi- and cross-), and interconnectivity contribute to synthesis.

SYSTEMS

An identified system is an orderly combination or arrangement of particulars, parts, or elements working together to produce some supposedly desired results as a whole, especially such combinations according to some rational principle and methodic arrangement of parts. There are systems that are not necessarily organized in an orderly way based on present knowledge. Some systems are chaotic-weather systems, turbulent flow systems, ecological systems. As more is learned about these systems that appear to be unorganized or at least not organized, predictive models are improved.

Synthesis is a process for which the system is a product, and can result in imaginary, natural, or artificial. Very few systems encountered in the real world are deterministic. The synthesis effort, usually of an artificial system, distinguishes a systems engineer from a systems analyst. The systems engineer usually has an objective of optimization of performance.

A constructed physical system is an example of synthesis. Some insight into the system might be provided by a study of biological systems, which until recently occurred without the intervention of human effort. The human body system consists of several (sub) systems: circulatory, digestive, endocrine, musculary, reproductive, urinary, and each of these have sub-systems or parts. Continued study contributes knowledge to better understand not only each system by itself, but the relationship of each to the other.Regardless, "one of the most pervasive and least well-defined concepts in modern intellectual thought is the notion of a system," according to Casti (1979). Theories of large scale systems are needed to provide a base for designed systems, systems of economics, politics and law, energy, education, environment, and transportation. Although considered as systems, these systems are parts of even larger systems.

Casti (1979) discusses internal and external descriptions of a system; internal and external constraints on a system; connectivity and simplicial complexes (linked through shared faces or common vertices); complexity (mathematical structure of irreducible component subsystems); description of how the components are linked to form a system; reduction of complexity by grouping subsystems; dimensionality, hierarchy, length and strength of connective chains; static and dynamic complexity; and stability (the ability of something to react to changes in the environment without failure or damage).

As one considers static complexity (presumably less complicated than dynamic stability), several aspects of connectivity must be considered, including, among others,

» Hierarchial structure

» Connective pattern

» Variety of components

» Strength of interactions

Complexity involves notions of nonintuitive system behavior, patterns of connection among subsystems such that prediction of system behavior is difficult without substantial analysis or computation, decision-making structures that make the effects of individual choices difficult to evaluate, and so on." (Casti, 1971) He presents six steps to take the complexity of a system into account. In these steps he considers an artificial system as a mathematical representation.

1. Attempt to find a reasonable analogy between a natural system and an artificial system
2. Compute various transformation semigroups for a given system, taking into account its past history and different possible designs
3. Compute the design complexities of the semigroups
4. Compute the rational level of contact function associated to the semigroup of transformations
5. Compute the control complexity of this contact function
6. Compute those contact functions and complexities that yield stable configurations for particular designs.

Hall and Fagen in Buckley (1968), list the following explanations of the words used to define a system:

» A system is "a set of objects together with relationships between objects and between their attributes."

» Objects are simply "the parts or components of the system."

» Attributes are "properties of objects"

» Relationships are those things that "tie the system together."

Singh (1987) distinguishes systems, systems science, and systems engineering. "A system is a complex entity formed of many, often diverse, interrelating elements subject to a common plan or serving a common, occasionally unknown, purpose." p. 4766. Systems science is "the branch of knowledge concerning systems and their behavior, properties and responses." p. 4810. Systems engineering refers to "the management, direction, control and technical effort applied to system design to ensure or maintain integrity and integration as related to design configuration, reliability, and performance in terms of efficiency and effectiveness." p. 4766.

Control, input-output information, provides a method of relating parts of the system. Comparison of control systems in engineering and biology is provided in Matrix 7-1, as developed by a biologist. An engineer involved with systems would probably disagree on some of the characterization of simplicity, but more importantly the biologist recognizes the important emphasis on synthesis by the engineer.

Although analysis appears to be the main activity in the study of biological systems, analysis practically always follows some kind of synthesis in natural, unnatural and artificial (humanmade), and human (social) systems as well. Without synthesis analysis could lead to a dead end and be less interesting. Through insight, creative imagination, and intuitive thinking, a person or group of persons can provide ideas, equations, theories, that will keep hundreds of

Matrix 7-1. Comparison of Control Systems in Engineering and Biology

	Physical	Biological
System control	Lumped	Distributive
Organization	Simple	Complex
Dimensionality	Low	High
Linearity	Linear	Nonlinearity
Time dependence	Invariant	Variant
Methodology	**Synthesis/Analysis**	**Analysis/Synthesis**
Nature	Goal seeking by deterministic and stochastic processes	Goal seeking by evolution and adaption
Interaction of subsystems	Fixed	Varying
Redundancy in design	None	Much

* Bold print added for emphasis

Source: Berlinski (1976)

people occupied in trying to prove or disprove that proposed relationship. Einstein did little laboratory work himself, but his paper on relativity continues to challenge others after nearly 100 years. Challenging proposals such as those by Einstein are often the result of synthesis; those that work to better understand the challenging proposals engage in analysis many times over.

CREATIVITY

Creativity is a desired activity that is important in analysis and synthesis, in the former case primarily of something that is, and in the latter case primarily something that has been visualized or made. Creativity is the process of bringing something new, creations, into being, usually by 'combining previously unrelated structures in such a way that you get more out of the emergent whole than you put in.' The four stages that characterize creative thinking are preparation (of the mind), incubation (thinking), illumination (analysis-synthesis), and verification (does it work). (Hall, 1989)

The criteria of creativity are given by Dutton and Krausz (1981):

» Created outcomes have intelligible structures that are irreducible

» Structures of created outcome are unpredictable

» Structures of created outcomes are inherently and usually instrumentally valuable (generic in nature?)

» Acts that lead to created outcomes include an element of spontaneity so that although they are directed and are controlled, they are discontinuous.

The pattern of creative thought seems to contain two kinds of thinking: realistic (oriented to external stimuli) and autistic (inner, subjective, emotion). (Hall, 1989)

CREATIVE THINKERS

Creative thinkers are characterized by the following attributes:

» Have curiosity and knowledge of many subjects beyond their identified interest at the time

» Make connections between knowledge in different subjects, often over looked by others

» Have an interest in not one but several possible outcomes

» Have ability to abstract a formerly obscure aspect or view from relevant data

CREATIVE WORK AND THE USER

> "Creativity is not simply originality and unlimited freedom. There is much more to it than that. Creativity also imposes restrictions. While it uses methods other than those of ordinary thinking, it must not be in disagreement with ordinary thinking—or rather, it must be something that sooner or later, ordinary thinking will understand, accept, and appreciate. Otherwise the result would be bizarre, not creative."
>
> *(Arieti, 1976)*

This thought parallels the need for the results of synthesis to meet more than the criteria of analysis, but the needs of potential users. Any creative product is considered from two points of view: as a unity itself that enlarges the understanding of the universe, and as a part of the culture that enriches and enlarges human thought.

DESIGN

The central theme of engineering is design. This statement does not mean that everyone 'designs' in the usual sense of putting the final product on paper, but that nearly everyone contributes to the design process. Design is a process and incorporates input from numerous people. The engineer in research and the engineer in sales should see their task as a part of the design process and be considered part of the design team. The salesperson should provide information on their products and be part of the design process. A role for engineers is to design systems, to synthesize systems utilizing subsystems and components and materials. The key words are 'design' and 'system.' Design is too often looked at as making a sketch or drawing. Design must consider the larger system involved. Guy (1987) states that "the essence of engineering is not design, it is engineering judgment." This view is substantiated by Hazelrigg (1991) who maintains that engineering is not problem solving, although solving problems is involved, but is decision-making. Decision-making means that there must be choices, and in making choices information is needed so risks and benefits can be considered in making judgments. Developing a design requires numerous judgments based on the problem. Many designs do not see the light of day because of their weakness. Yet the public doesn't see the work involved in these unsuccessful designs and some people have the impression that engineers build devices without considering the public needs, wants, and welfare. Unfortunately this was too often true in the past when the larger system in which the device was to be used was not considered for public use. Now fortified with a supporting public opinion, some engineers have become the nation's leading exponents of environmental maintenance or improvement.

AS A PART OF GLOBAL SYSTEMS

In discussions with engineering students from foreign countries, particularly developing countries, it is obvious that many have difficulty seeing an appropriate role for them upon returning to their country that is devoid of manufacturing facilities to make and to produce things. Many of the manufactured components for a system in a developing country may be imported, but these components need to be used in a system appropriate for the country. Adapting the broader view of considering the entire system, of selecting or synthesizing a system, is as challenging, can be as creative, and certainly as useful to a developing country as the manufacture of parts of the device, machine, or system in the developed country. With the move to global enterprises, a developing country could become a participant in design, development, or manufacture of a part of a system.

U. S. STUDENTS, ANALYSIS AND SYNTHESIS

The role of systems building, synthesizing, is not adequately presented to undergraduate students in the U. S. Students recognize their strength in analysis and their weakness in synthesis. That strength in analysis is needed and must not be destroyed or depreciated, but the weakness in synthesis must be corrected, and can be done so within existing courses. Some colleges and universities include instruction and experience related to synthesis in architecture and engineering through courses, laboratories, internships, design assignments, industry shadow programs, industry seminars, and many more. Accreditation procedures encourage these approaches but haven't been as effective as needed.

FUTURE DIRECTIONS

The National Research Council (1987) study, "Directions of Engineering Research," identifies the future engineering research as falling into the following systems:

» Bioengineering Systems

» Construction and Structural Design Systems

» Energy, Materials, and Environmental Systems

» Information, Communications, Computational and Control Systems

» Manufacturing Systems

» Materials Systems

» Transportation Systems

By classifying engineering research into these categories there is a greater recognition of synthesis than any recent study. Usually such studies have been made on the basis of disciplines.

AN ACTIVITY IN SYNTHESIS

Design is an activity in synthesis. Until recently engineering design consisted of making a sketch or drawing, often with precise details and instructions for making an artifact or device. Those designs usually considered the functioning of the device itself without more than a cursory consideration of the energy required and impacts on the environment and human concerns. Those designs were based on primary factors. Today, and in the future, designs must consider secondary, tertiary and beyond impacts, negative and positive. Secondary factors may become primary factors as their importance is recognized. Safety of toys, tools, ladders, appliances has moved to the forefront as a factor in design. Fuel economy for an automobile is no longer a secondary factor but a primary factor in design. Pollution control was probably a tertiary factor in design of an automobile and is now a primary factor. As more factors are considered and included in the design of an artifact a greater effort in synthesis in involved.

COSTS AND BENEFITS

In economic studies it is often difficult, if not impossible, to appropriately determine and incorporate the costs and benefits of secondary, tertiary, and beyond factors. As an example, consider the difficulty of comparing the cost of electricity produced from different sources such as fossil fuel, hydroelectric, solar, and nuclear power. The first or primary costs would be straight forward as land would be purchased, structures built for the three types of units, and a maintenance and operating costs could be included. It is difficult to calculate and assign the secondary costs such as impact on future food supply, change in the natural environment, radiation effects (not limited to nuclear plants); or the tertiary costs of the research and development programs that were funded to provide a workable unit; or the quartiary costs of providing the education to prepare personnel. The situation becomes even more complicated as some people but not others see benefits such as a lake for recreation when using the hydroelectric approach. These approaches require a greater synthesis than in the past.

In energy studies consider the difficulty of determining the net amount of energy produced and the comparative cost of producing gasoline from shale. For transportation systems, to what extent do you include the energy not only for fueling the vehicle on the highway but the energy for building and maintaining the highway. An additional step could be the energy used in developing various materials and designs, with an additional step involved in energy used in research, including the costs for results obtained for those materials developed but not used. These and similar considerations will require a greater synthesis than in the past. The economic factors in a competitive system greatly influence the performance and choice of systems, but these must be a part of the synthesis approach. The legal costs in many systems is becoming an increasingly important part of the costs.

A Whole Environment Evaluation Synthesizer (WEES) is proposed for studying the three phases of gas, liquids, and solids over a period of time. The evaluation of environmental changes considering various energy inputs related to the prebiotic environment is made for primary, secondary, and tertiary effects.

Discussions regarding costs and benefits are usually partisan in which decisions are based on "who pays the costs" and "who gets the benefits." Using an activity-based accounting approach, special interests must be assessed and decisions made on the basis of a larger system than might be under discussion. Public policy and law are often developed to protect the larger system (state or national government) from abuses by smaller parts of the system, as exemplified by polluting automobiles (operated by individuals) that damage the atmosphere used by everyone in many situations the 'smaller or minor' participants must be protected from the 'large or financially-strong' stake-holders. Political discussions are often confused by one group that wants to obtain the benefits but have another group pay the costs.

DESIGN AND GENERALIZATION

The verse presented by R. Buckminster Fuller emphasizes the point, however, that in spite of the need to generalize and take a global view, one must get specific when making the design.

> "Man cannot design
> A generalized anything
> He can only embody
> The eternal design's generalized principles
> In special case embodiments
> Of those generalized principles."
>
> *Fuller, R. B. (1972)*

All engineering is heuristic according to Koen (1985) who explains the concept by listing and describing synonyms for heuristic, such as rules of thumb, orders of magnitude, intuition, factors of safety, engineering judgment, engineer's attitude toward work, or, if in France, te pif (the nose). He describes engineering as a problem-solving, goal-directed and needs-fulfillment activity. The heuristic approach is used to describe the plausible, even if fallible, while keeping risk within acceptable limits, to cause the best change in a poorly understood situation with available material and financial resources. Thus, the engineering method is the use of engineering heuristics. ". . .that general heuristic rules—use of analogy, appeal to symmetry, examination of limiting conditions, the visualization of the solution" is used by intuiters (but if used frequently by intuiters will be anything but support to intuitive thinking) but that does not necessarily lead to the correct answer. (Bruner, 1963) The synthesis approach incorporates the heuristic procedures to come up with well-founded opinions and information.

DOXA

Doxa is the name given to well-founded opinions supported by evidence and reason; the right opinion as contrast to 'knowledge' which has the connotation of 'truth.' (Adler, 1993) Doxa is knowledge in the sphere of doubt, not knowledge beyond a shadow of doubt. Thus, knowledge can be modified or changed at some future date. He writes, hopefully, "the illusion of episteme (truth) will be replaced by 'a sober respect for testable doxa as the only grade of organized knowledge that is achievable within philosophy or science.' " Engineering and the professions and applied areas, strong in synthesis, differ from science in that decision-making and actions are made on the basis of partial knowledge identified as doxa. Doxa is modified based on the results of experience. Many questions in engineering cannot be answered with the certainty of many science questions, represented by episteme. When considering the 'truth' or 'proof,' doxa represents a phenomenon that is not as close to being provable as apodictic. Apodictic phenomenon can be clearly proven or demonstrated, or is incontestable in terms of the state of knowledge.

Matrix 7-2. Inquiring Systems to Seek Information Based on Philosophical Approaches

Philosophical Basis	Description, Representation	Problem Identification and Use
Leibnizian	Start with set of truths; symbolic; mathematical; base on laws of science.	Well-structured, defined problems for which analytic solution exists.
Lockean	Empirically based system; consensual; use traditional databases, accounting systems, statistics.`	Model is known, have a priori relations; well-structured problem with consensus of opinion.
Kantian	Multimodel, synthetic system; at least two representations of each problem or task; partly scientific, partly empirical.	Moderately ill-structured problem or task.
Hegelian	Present two or more conflicting opinions; can be preexisting opinion or created for consideration.	Ill-structured problem; gain better view for understanding conflict.
Singerian	Provides for continued learning through feedback; transform structured into ill-structured problem or vice-versa.	Ill-structured problem; Compare results with various approaches and select best.

Referred to as Churchman's Concept.
Compiled from Churchman (1971). Also presented in Singh (1987).

PROBLEM SOLVING

The scientific method is widely heralded by many as an approach to problem solving, that follows problem identification (Matrix 7-2) often overlooking the fact that the scientific method applies to science and the scientific enterprise. The scientific method is usually used to solve questions self-identified and to obtain information judged by and used by peers. To illustrate the similarities and dissimilarities, Matrix 7-3 compares the scientific method and the engineering method. In a way these are methods of problem-solving, with modifications to fit the activity, and can be applied to every day situations. Some of the steps are taken without deliberate thinking, that is, they comenaturally. Other steps in the process are taken deliberately to assure a productive result. The major difference between the two methods is the analytic approach for the scientific method and the synthesis-analysis approach for the engineering method. The involvement of the engineer in problem identification and the involvement of the potential user in identifying the problem (to be defined in more detail by the problem solvers) and in determining alternative solution, and in evaluation of the product or process is inherent in the synthesis approach. The management of the technological process is engineering (Guy, 1987), in which economics is a major element.

Matrix 7-3. Comparison of Scientific Method and Engineering Method

Scientific Method	Engineering Method
Question Identification	Problem Identification
Existing Knowledge-Analysis	Existing Knowledge and Experience-Analysis
Scientific Curiosity	Creative Approach
Hypothesis	Conceptualization
Analysis	**Synthesis**
Proof	Evaluation of Alternatives
Publication	Decision/Implementation

* Bold print added for emphasis

Source: Hall (1994)

INDIVIDUAL AND GROUP ACTIVITIES

Idea generation, problem-identification and problem-solving can benefit from team or group efforts in addition to individual activity. Idea generation is enhanced by team member interaction (Image 7-3). Problem solving performance is improved by cooperative efforts, demonstrated quantitatively for computer type problems (Hall, 1989). The speedup of idea generation is about two orders of magnitude greater for cooperating agents than that obtained for the non-cooperating case. "The major difference between cooperating and non-cooperating agents is that hints effectively reduce the size of search space by focusing the agents on much more plausible courses of action." Also, an agent searching one part of the space may find a hint useful to another agent in a

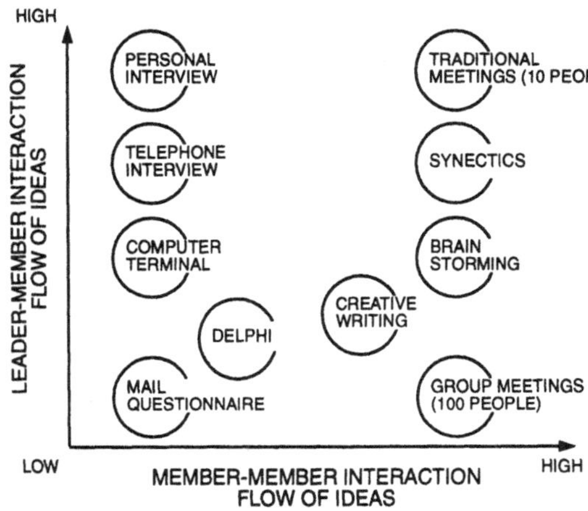

Image 7-3. Interaction Effects on Techniques for Getting Ideas

defined part of the space. Thus, the productivity is greater than that from individuals operating separately.

As scientific and technological efforts increase more people get involved in collaborative efforts involving two people, groups or teams, whether called interdisciplinary, multidisciplinary, or cross-disciplinary, or in the efforts to design, develop and manufacture devices. People in many fields of endeavor in which individual activity has previously predominated have spoken and written in opposition of the trend to interdisciplinary and multidisciplinary activities, particularly in research partnerships. These people argue that the creative abilities of the individual are diminished by such efforts. They argue this point not only for their fields that have been predominantly individual, but also for other fields such as engineering. Steinbeck (1952) writes in 'East of Eden,'

> "Our species is the only creative species, and it has only one creative instru ment, the individual mind and spirit. Nothing was ever created by two people. There are no good collaborations, whether in music, in art, in poetry, in mathematics, in philosophy. Once the miracle creation has taken place, the group can build and extend it, but the group never invents anything. The preciousness lies in the lonely mind of a person."

CREATIVITY STIMULATED BY OTHERS

The view by Steinbeck is accepted by many but fails to recognize the importance of the stimulation that comes from others in producing a creative thought or idea particularly of an unnatural or constructed system. Ideas are not produced in vacuum. Even the creative writing of Steinbeck was stimulated by external influences, albeit that those were less organized and more difficult to identify than those ideas that occur by an individual working as a member of a team of people. As a result, papers and inventions will be credited to two or more people without knowing exactly which person's brain produced the idea, but recognizing that the idea might not have developed by either person working independently. Two recognized painters for one painting is rare, but not unknown. In the Seattle Museum of Art is a painting by Gilbert(b. 1943) and George (b. 1942) titled, "Coloured Shouting," 1982. It is the rare individual who works entirely on his/her own. Autobiographies of people who discuss their productivity/innovation invariably identifies some external influence that stimulated an internal creative idea, Matrix 7-4.

GROUP AND CREATIVITY

Gordon (1961) warns us that there is no such thing as a group mind. But this doesn't mean that creativity is necessarily dampened by group or team activities. Group activities, in which each person, or each smaller group, contribute to an overall objective, are common. The leadership of a team or group has much to do with the creativity of the activity. Children, along with their peers, prepare a poster to focus attention or show concern for an issue, to honor a desirable activity, to encourage mankind to live safely in harmony. People contribute a 'piece' to a quilt, individuals contribute time and money to a larger cause than their own, to assist their fellow humans in time of sickness and stress, and help people to obtain or keep their freedom. A synthesis of ideas, parts, pieces, and activities occurs to reach a common objective of individual parts or participants.

GROUPTHINK

When participation by teams and groups are involved, a major concern is the extent to which each participant provides or is provided the opportunity to participate constructively. Groupthink is a term used to describe an undesirable approach or practice to limiting participation. Groupthink can be minimized by having a leader that does not have an authoritarian style and who is not an advocate, who permits and encourages all people to participate, and who aims for a high quality consensus not forced excessively by time constraints or pre-determined views.

Matrix 7-4. External Influences that Stimulated Creativity of People

Person	Creative Result	Stimulating Activity-Environment
Georges de Mertral, 1957	Velcro®	Curiosity-study burdock seeds under microscope (invention)
Art Fry, 1977	Sticky tabs	Need to mark pages in hymnal with out damaging pages (invention)
Marc Gregoire, 1954	Non-stick pan	While perfecting a fishing rod (invention)
John Loud, 1888	Ball point pen	Search for writing instrument to avoid ink bottle and cleaning pen
William Mason, 1925	Masonite	Search for way to use wood chips and sawdust, leaky steam valve cooked
Louis Pasteur, 1880s	Vaccination	Search for way to prevent earth from anthrax bacteria
Roy J. Pluncket, 1938	Teflon	Keen observation of laboratory accident (discovery)
Erno Rubic, 1979	Rubic cube	Search for challenging game (invention)
James Watt, 1765	Improved Newcomen engine	Search for way to improve efficiency of steam engine adding a cylinder

References:
Derry, T. K., and Williams, T. I. (1961)
Messadie, Gerald (1991)

QUALITATIVE PHYSICS

The synthesis approach is represented by what some engineers call qualitative physics in which 'back of the envelop' or 'seat of the pants' methods are converted into an expert system using the computer. 'Back of the envelop' sketches, or rough calculations, for example, of the amount of dirt to be removed for an excavation, often provide order of magnitude representations suitable for many considerations. This approach can be just as successful as Lincoln's perceived 'back of the envelop' Gettysburg address. Qualitative notation is used in expert systems to represent physical variables, otherwise represented by quantitative methods, such as differential equations. (Stix, 1991) The technique has been used for representing repair and adjustments procedure, as a fault finder for a jet engine, and for design tools and for instructional purposes.

ENGINEERING TEAMS

Creative approaches in engineering are fed by associating with others—both the producers of knowledge and those with experience that might be the basis of a "thing" and the potential users of the "thing." One definition of creativity is the ability to combine unrelated subjects to provide a new idea or approach. Thus, one needs to be continually open to producers and sources of knowledge and experience and consider possible uses to perceived or real questions or problems. Partnerships among organizations and institutions is a mechanism to promote the participation of their people to work together. A clear statement on the relationship of an individual to a team is given by Moranian (1963):

> " ...the objectives of the [engineering] research project must be achieved through individual efforts of team members... being a member of a team is not necessarily bringing about less individual creativity... "

The appropriate use of individual and group activities is important. The classic example is solving a crossword puzzle and making a crossword puzzle. The first is most easily done by a group; the second is most easily done by an individual. (Connolly, 1983)

GROUPWARE

Groupware is an emerging field of computer software that provides a tool for two or more people to work together interactively at diverse locations running diverse types of computers to carry out a common task. The software in itself is not necessarily creative, but assists in the 'creating' activity. Data and information are accumulated from a number of sources. Each person participates in adding to, modifying, or removing information, each from his/her terminal, subject to the approval of the others. A common task could be a make a design, provide recommendations, write a report, or optimize a function. The product is the result of all members of the group. Groupware provides a valuable assist for synthesis of ideas and information from several people at different locations to work toward a common goal.

QUALITY/QUANTITY

People who emphasize synthesis are often confronted with considering qualitative factors to a greater extent than those who emphasize analysis. Quality assurance (QA) is probably more of a quantity analysis than generally realized, and is acceptable in the sense that it should be understood that QA by itself does not present the total picture.

Quality is a non-quantifiable measure of an idea, a thing, or an activity, although elements may be quantifiable. Thus, an object might be identified by several quantifiable elements such as shape, color, smoothness, about which we would all agree, but these elements can be combined in several different ways, and the observers may reasonably evaluate quality of the objects in different ways, and no one quality can represent the object. Quality is made of quantifiable or a combination of quantifiable and qualifiable (non-quantifiable) data and information. The result could well be a qualification of a product that is not agreed to by many, even though there would be agreement of the quantified elements, because of the manner the quantifiable elements are combined.

Quality includes a diversity of thought, a uniqueness of perspective that enriches the intellectual climate. Generally when a thing can be quantified, it is no longer in the qualitative category notwithstanding that several quantitative measures may identify aspects of an item, and various observers may put different weights on measures of importance, thereby resulting in qualitative judgments.

The thinker or participant making a judgment is in a position to make qualitative judgments based on a combination of several quantitative factors. Thus, for example, through qualitative judgments, a person would prefer a particular thing (food, house, beverage) even though numerous quantitative measures are available. A particular food has a color, odor, taste, feel, all of which can be quantified. Yet different people would qualitatively prefer different items based on a synthesis of or combination of particular characteristics.

EVALUATION OF SYNTHESIS

The result of synthesis needs to be subjected to validation. A major strength of the analysis approach is the ability of others to evaluate the results through repetition of the process and duplication of results. With synthesis, however, as one looks outwardly, to a larger system, the process of analysis is only partially applicable. Often the systems are so large and complicated that they cannot be duplicated by others, or only one such system is available, often not describable. There is room for "mischief" and unreasonable "dreaming" such that the system cannot be duplicated by others. Thus, the validity of results of synthesis is difficult to verify, particularly when conventional wisdom is challenged. Projecting from one level to the next level (of size, of complexity, of hierarchy, of task to be performed, etc.) minimizes the potential for error as contrasted to skipping several levels.

If synthesis could be completely evaluated by analytic means, the activity under consideration would not be synthesis, but would be analysis by definition. Certainly synthesis must stand up under analysis, but that is not sufficient to determine the value of the results of synthesis, which is to meet greater needs and to serve larger purposes, beyond analysis. Synthesis cannot violate natural laws, although those laws might be improperly conceived, so must be reconsidered if the information demonstrates such a need. Synthesists in the professions need to do their work within the code of ethics established by peers and published by professional societies, another point for evaluation. A principal need of the result of synthesis is to serve the (potential) user, or consumer, of the information, product (artifact), or process. A so-called beautiful piece of art, a poem or novel, an appliance, an automobile has to have the acceptance of a viewer, reader, or user to be a good synthesis, while at the same time meeting the criticism of analysis. Esthetics is an important part of evaluating the results of synthesis. The results of synthesis must meet the laws established by government, as well as not violate natural laws, and should be done in an ethical manner.

Analysis is strong in quantification, using quantity factors to describe the phenomenon, findings, and results. Synthesis provides an opportunity to also include qualification or quality factors. There is a continuous attempt to quantify qualitative measures. Yet when quality is quantified, those quality factors are no longer such, but are quantity factors. Quality may be described by several different quantity factors. Different evaluators may place different weights or value on the combination of quantitative factors, thus preserving the qualitative nature of the evaluation.

The approach of synthesis must not be confused with magic, myth, occult, or witchcraft. One approach is to avoid expanding the system under consideration (in the real world) beyond realistic boundaries. At least a vision of the potential should be described in words, if not in numbers, perhaps within an order of magnitude. How to determine realistic boundaries by avoiding unnecessary restriction and without dampening creativity, but still encouraging an expanded system or view, requires judgment and practice.

INVOLVEMENT OF PEERS

Traditionally peers have been valuable participants in determining the value of work done or anticipated in science. Peers are useful in determining the merit of engineering work anticipated, but should be considered differently than for science. The peers for considering engineering synthesis, as contrasted to science approaches, should include people in 'user' category (clients and consumers) as well as the 'producers'; peers should consider not only individual evaluation responses, but also use team, group or panel meetings; objectives of the work should be described by a level in the hierarchy above the 'producer'.

SUMMARY

Many natural laws are applicable and useful but fall short in predicting the function of artificial systems. The emergent quality and properties of unnatural or artificial systems need to be investigated because the information about the parts themselves is not adequate. As more information is obtained the emergent quality of a substance increases and more emergent qualities can be predicted.

Synthesis is a natural process, as in photosynthesis, and is an unnatural or artificial process as in making a device or building. New inanimate systems are synthetical, a result of intentionality, either based on innate properties or to meet user needs. Synthesis is the process of which the system is the product. Although a clear, comprehensive, generic description of a system is still being sought, one may consider a system as a whole made up of parts, the parts within the whole are adopted to achieve the purpose of the whole, with the whole setting the purpose.

The major change in design today as compared to a couple of decades ago is that previously the primary factors set the design, whereas secondary and tertiary factors are now involved as well as more emphasis on legal, ethical, esthetic, and environmental factors. Design is the central theme in engineering and architecture, and design is primarily a synthetic process, incorporating appropriate analysis. Although there is an overall drive for generalization with synthesis, one has to be very specific in the design.

People who emphasize synthesis in their activities deal with qualitative values to a greater extent than those involved in analysis. Quality is a non-quantifiable measure, usually containing quantifiable elements. Several quantifiable elements may go into a qualitative judgment. Different people could combine these quantitative elements in different ways to get different answers often causing controversy on a decision.

Part 8.
Syntax and Summary

There is *a posteriori* evidence that synthesis and synthetic thinking are becoming more important in the world, both philosophically and technologically. Postmodern eras in art and architecture incorporate a marked increase in synthesis. Engineering and medicine are increasingly using synthesis methods to better serve their customers, consumers, and clients. Books have been written in practically all fields emphasizing the importance of synthesis. Several scholarly publications are now available, not only including the word synthesis in their titles, but include many cross-disciplinary titles covering multitudes of subjects. Some fields, such as chemistry, have developed entire new branches such as synthetic chemistry, often outside their original disciplinary home. With an increase in the number, size, and complexity of built-systems and artifacts, synthetic thinking has been involved in the design, building, and understanding of these artifacts. Teams of individuals with different specialties with appropriate leadership are required to synthesize the various systems. Integration of ideas, organizations, materials, and information is a major endeavor, with integration being a step toward synthesis and synthesis as a step toward unity. Systems approaches and studies in progress provide a strong precursor for synthesis.

RECOGNITION OF NEED FOR SYNTHESIS

Many leaders are speaking and writing about the need to recognize the importance of synthesis, particularly in engineering, where the education and training are believed to lack adequate coverage of synthesis. World prizes, such as the Nobel Prize, Draper Award, Deming Prize, and National Medal of Technology (U. S.) are being awarded to a greater extent to people who contribute as members of a team or group. Also recognized are individuals in leadership positions who are involved in synthesis as opposed to analysis, but are more likely utilize to synthesis-analysis efforts.

INFRASTRUCTURE IN PLACE TO ADVANCE SYNTHESIS

A major development of national and worldwide importance occurs on the foundation of previous developments and existing thinking is supported by a major infrastructure. There is a physical infrastructure that supports construction, transportation, trade and communications. There is an intellectual infrastructure, as well as a physical infrastructure, that supports synthesis. The building blocks of synthesis, incorporating elements of analysis, are represented in Image 8-1.

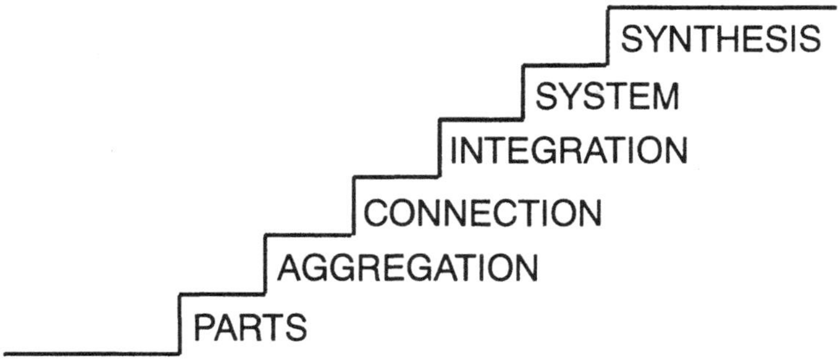

Image 8-1. Steps to Synthesis

There are many indications that the infrastructure is in place and is becoming strengthened to sustain synthesis as a major way of thinking and doing in the U. S., as listed below:

» Precursors for a strong synthesis are now well-developed and in place, principally a well-developed capacity for analysis, plus an increasing realization of the importance of synthesis in the rapidly developing global enterprises,

» Developments in information storage and handling have put the world's population in closer touch, making people knowledgeable of larger systems,

» Some disciplines, such as chemistry, now have an active and strong synthetic group, with the synthetic group building on the work of analytic groups, not replacing it,

» People in the arts (painting, sculpture, theatre, literature) have been utilizing and integrating the latest technologies into their work,

» The professions of architecture and engineering have grown in importance and recognition, and through the design process synthesize information from social sciences and physical sciences to provide constructed devices and systems to meet user needs,

» Although the word photosynthesis has been in use for many years, other words related to using synthesis include biosynthesis (production of chemical by a living organism); chemosynthesis (synthesis of an organic compound by energy derived from chemical reaction); and polysynthesis (united many parts into one),

» The major physical and social systems result from synthesis of smaller to larger systems. The physical and social systems are developing stronger interrelations among communications, energy, transportation, environment, economic, education, health, and political fields,

» Engineering grew out of the crafts, as an art, then more recently incorporated the sciences, and will now integrate these two areas and emphasize synthesis,

» Efforts by many organizations to "plan" for the future, ten years or more, are primarily intuitive processes, incorporating the synthesis of ideas and perceptions,

» Consideration of major subjects will be in terms of systems, meaning more of a synthesis will be involved, and

» Educators and researchers are incorporating more activities relating to synthesis.

EDUCATIONAL REFORM

Walters (1988) writes "that educational reform occurs in a profession when newly educated members find themselves unprepared to fulfill the roles for which they were educated." We can add to this statement that if the "users" of the products of education and training are unsatisfied with the product, reform is accelerated. The suggestions of reform leaders are monitored by gate-keepers, those in the power of authority in influential positions in the profession. For education, reforms are represented primarily by changes in courses and curriculum, usually based on experiments or trials at a limited number of institutions, and those with the most convincing results are adopted by other institutions and by professional organizations. In engineering several trials are being conducted on synthesis programs in education and research, primarily at universities in terms of Federal support activities. Engineering Research Centers (ERC), of which there are now twenty; and Engineering Education Coalitions (EEC), of which there are six; and Industry/University Cooperative Research Centers (I/UCRC), of which there are now fifty, have been partially funded by the National Science Foundation to improve engineering education, research, and practice. The I/UCRCs have been a forerunner of the expanded activities of cooperative research between higher education and industry sectors and their success led to greater support, both from government (local, state, and federal) and industry, as a means of improving education and assisting the economy in the short and medium term. Many other agencies and several countries have used the ERCs as a model to strengthen engineering.

The Engineering Research Centers have in common:

» research and education focused on technological advances critical to increasing the competitiveness of U.S. industry;

» cross-disciplinary teams emphasizing an integrated systems view of the field,

» active involvement of industry in research and education;

» exposure to undergraduate and graduate students to team research as well as to industrial perspectives on research, design, and manufacturing; and

» a commitment to technology and technological information transfer.

The Engineering Education Coalitions have in common:

» a vision and well-formulated plans for comprehensive restructuring and reshaping of the undergraduate engineering learning experience involving the breadth of engineering and all levels of undergraduate engineering education;

» an agent of active inter-institutional collaboration with significant intellectual exchange and resource linkages;

» tangible resources that reflect the unique interests and capabilities of the institutions;

» well-developed plans for integrating the proposed activities into the academic programs of the institutions; and

» baccalaureate-producing institutions with activities that exhibit a broad perspective of the engineering education system.

The Industry/University Cooperative Research Centers have in common:

» participation by more than one industry to help assure that results support more than one company;

» relationships develop between students and faculty and the industry in problem-identification and problem-solving; exchange of personnel among participating groups;

» after an original investment in start-up by government agencies, the financial investment is by the users or benefactors of the results of the center; over a 15-year period, the industry and other external support has been fifteen times the government investment; and

» work on interdisciplinary research directed toward meeting needs of in dustry

PUBLICATION AND SYNTHESIS

Publication (thesis, theses, treatises) of thoughts is one way of advancing ideas and development of things. Authorship has become an important criteria for evaluating the performance and productivity of scholars, particularly in research and academics. Publications of a scholar are considered on the basis of the reputation of the journal, the title, the number of pages, often with little respect to the contents of the articles.

The majority of people evaluating publications are from the predominately analytical disciplines, most of whom do not rate highly those articles that are primarily synthesis. Analysts usually write papers to convince their peers who are analysts, and don't look kindly on synthesis papers. This is particularly true for engineers where publications are judged by committees consisting predominately of scientists within the academic community. The publications of engineers must convince not only the analysts of their value, but also the users who are more likely to be synthesists. Users or potential users are usually overlooked in the review process, or their views are discounted. Articles by analysts usually cover a subject in great depth, in which a laboratory experiment can be a complete article, with little consideration of the larger picture.

Veterinarians, medical doctors, engineers, architects, agriculturalists, all of which have a science base, utilize pieces of information from several sources to put together a larger picture (synthesis).

The synthesis-type articles usually have more than one author, which raises the question by the members of evaluating committees, 'Can this person do independent thinking?' Or if the article is written by one author, 'Can this person work with others?' A person in academia who writes books comes in for special criticism, particularly if a book becomes popular and returns substantial royalties to the author. Isaac Asimov comes to mind, a person who remained at the associate professor level, while informing millions of people through hundreds of books on science and related topics. In the business field Peter Drucker is a master synthesist; in philosophy, Mortimer Adler; in engineering, W. Edwards Deming, and Samuel Florman; in education, Jacques Barzun; in general, R. Buckminster Fuller; and in novels, James Michener. There are many more. Yet many books, like comprehensive review articles, represent a synthesis of information that is of true educational value, and can do much for the efficient generation and use of knowledge and information. More people in academics need to understand this relationship, not only in their own teaching, but in their criticism and evaluation of colleagues. Until synthesis takes its rightful place, we will be limiting important contributions to meet the needs of society.

Although titles to articles and papers are heavily slanted to analysis, a goodly number now have titles including the word synthesis. However, many articles that imply a synthesis are based on articles published as analysis, a kind of state-of-the art with very little synthesis involved, except after the fact.

More effective articles will more likely result by using a synthesis approach from the beginning of the thought process or research project rather than after. The situation is similar to an engineer being asked to make sense from insufficient data obtained when the engineer was not brought in on the instrumentation of the project. Those articles that cover the making of synthetic materials are excluded from this limitation.

Organizations must seek ways to reward those who produce based on synthesis, such as by leading in developing a vision and relating that vision to action; by encouraging and recognizing the value and importance of review articles, reviewed not only by analysts but also synthesists; for sharing with the employee the returns from products that generate revenue by an invention through license and royalty returns; and teachers who present their subject matter in context, demanding the rigor appropriate for synthesis as well as for analysis.

THE SYNTHESIS/SYNTHETIC APPROACH WOULD SERVE US BETTER

The combined analytic-synthetic approach instead of separate approaches could be used for all fields of endeavor, the arts and sciences and the professions and applied fields. The use of the analytic-synthetic approach (instead of the inductive-deductive approach by scientists) would put all fields on the same basis and tend to reduce the tension between the two cultures (arts and science) and encourage a fusion of the cultures. The synthesis approach will encourage development of a vision incorporating several approaches and provide several alternatives from which to select solutions. It will encourage an outward approach that is more likely to serve society, recognizing that analytical results are indispensable. Of particular value, synthetic approaches can help identify knowledge areas for which analytic work is needed and the results of imaginative thinking will feed back to stimulate those in analysis.

To accelerate the use of synthesis and to obtain the inherent advantages, synthesis should be incorporated as a part of instruction and research–as a way of thinking, acting, producing. Synthesis should be recognized as a scholarly activity and be incorporated into the life of academic institutions, particularly for the professions. Professional societies should make a place for 'synthesis' in its committees, programs, and publications.

SET OF RULES

A set of rules and methods is needed for synthesis as is available for analysis. Likewise, procedures are needed to evaluate synthetic systems, possibly based on present systems theory, hierarchies, and integration, or some new approach waiting for a creative mind. An attempt is made to describe a method of synthesis in Matrix 8-1.

Matrix 8-1. Method of Synthesis

1. Have a vision beyond the present system, subject, or thing that is a conceptualization of a larger whole, a more efficient or effective device or system that can be represented by an image, view, and concept, and expressed in words, numbers, and illustrations.
2. Develop a structural and/or functional framework to accommodate the vision in terms of words, numbers, and illustrations, considering the hierar chal relationships.
3. Identify connections and interrelations with external systems—economic, human, political, social—that might be involved, as is or as anticipated, particularly by users and potential users of the idea or product.
4. Identify connections and interrelations with internal systems or subsystems—economic, human, physical, political, social—involving control and operation of the present or potential product.
5. Evaluate the system on the basis of meeting a critical analysis of parts or elements and combinations thereof, by informed peers, of meeting the needs and wants of potential users of the system—economic, physical, political, social—and of the environment in which the system will operate or function, considering benefits and risks, such as health and safety.
6. Modify system on the basis of critical evaluation and creative thinking and realistic combination of present and anticipated connections, interrelations, and complexity/simplicity.
7. Repeat process until an optimum solution is obtained while considering various acceptable solutions.

Most of the emphasis in education and academic institutions has resulted in developing rules of thinking for analysis rather than synthesis. These rules have been developed over a long period of time based on logic (often mathematical), on arguments (syllogism-testing major premise by deductive reasoning), and by dialectic approach (arguments and counter arguments until resolution somewhere between). A special unique set of rules for synthesis is needed. Certainly it would seem that the rules for analysis could not be used productively by substituting the word inductive for deductive. Developing the rules for synthesis could be the basis of study for a dissertation, monograph, or treatise.

Elements of such rules are probably available now, but need to be brought together (synthesized) into a unity not now recognized. The study of systems, hierarchal arrangements, assembling, and design, all involve complexity and interconnectivity (connexity-Whitehead), have elements of a synthetic approach in philosophy. It would be desirable for such rules (statements) to cover both natural and artificial systems. A difficulty in combining both natural and artificial systems under the same rubric is how to account for the "end use" elements of artificial systems as applicable to natural (usually biological) systems.

A GENERALIZED APPROACH

Some natural systems assembly (self-assembly) is controlled by the impact of the elements, energy, and timing. For other natural systems, like plants and animals, we cannot attribute end use to a human need but perhaps to an evolutionary principle. For artificial systems conceived and built to meet a human need or want, end use can be an important driving force.

Although it is not clear that all systems are hierarchal (unless we define a system that way) the writings by Whitehead (1949) could be a start for developing a generalized philosophy of synthesis:

> "an 'abstractive hierarchy based on 'g'' where 'g' is a group of simple eternal objects, is a set of eternal objects which satisfy the following conditions:
>
> (i) the number of 'g' belong to it, and are the only simple eternal objects in the hierarchy,
> (ii) the components of any complex eternal object in the hierarchy are also members of the hierarchy, and
> (iii) any set of eternal objects belonging to the hierarchy, whether all of the same grade or whether differing among themselves as to grade, are jointly among the components or derivative components of at least one eternal object which also belongs to the hierarchy.
>
> It is to be noted that the components of an eternal object are necessarily of a lower grade of complexity than itself. Accordingly any member of such a hierarchy, which is of the first grade of complexity, can have as components only members of the group 'g'; and any member of the second grade can have as components only members of the first grade, and members of 'g'; and so on for the higher grades. The third condition to be satisfied by an abstractive hierarchy will be called the condition of connexity.
>
> ...What is meant by the notion of the connectedness of an actual occasion. This connectedness of an occasion is necessary for its synthetic unity and for its intelligibility. There is a connected hierarchy of concepts applicable to the occasion, including concepts of degrees of complexity."

The book gives additional discussion but it seems clear that hierarchy, complexity, and connectivity, are principal elements that need to be understood in synthesis (of systems). It seems that synthesis of artificial systems should also include control, one or several variables, redundancy, feedback, feed forward, and perhaps modeling.

SYNTHESIS AS A PATTERN OF NATURE

As more is learned about natural systems, that knowledge will be used not only to change those systems (genetic engineering, for example) but also to mimic for developing artificial systems (artifacts). People in the professions, people who design (in the broad sense), architects, engineers, and medical doctors, need to have a stronger understanding of the role of synthesis in their work. New fields are developing within disciplines, such as synthetic chemistry that require an entirely new approach to their work. The Nobel Award is now given to recognize the importance of synthesis (for CATScan, DNA, economics) in contrast to analysis. Some scientists will object to this approach in that their fields have more often been built on analysis, reductivism, and particularization rather than on synthesis, integration, and generalization. Then let us identify analysis as the scientific method and synthesis as the engineering method (such as genetic engineering by biologists). The incorporation of synthesis as a major approach represents a major paradigm shift.

ANALYSIS AND SYNTHESIS AS COMPLEMENTARY

In general, synthesis is a process of which the system is the product. Analysis and synthesis are often thought of as opposite terms. Rather, they should be thought of as complementary terms. Analysis can be done without much attention to synthesis, but synthesis usually depends on analysis.

The results of synthesis (synthesists) have probably done more to 'revolutionize' the technical world than the results of analysis (analysts). One must accept that the synthesists have higher risk and suffer more failures in meeting their revolutionary role because of speculation, intuition, and unpredictability in the process. Another thought is that analysis is more likely to lead to an epsilon (E) increase, a small change, in understanding, and the synthesis is more likely, if successful, to lead to a delta (Δ) increase, a large change.

Humans start out in life with an inborn curiosity directed toward analysis. Watch a small child explore by disassembling an item, and some breaking, while taking an item apart. Some synthesis is exhibited as children build with blocks or sand castles or items with erector sets or do artwork. Watch animals as they encounter a physical object, perhaps not an analysis, but as an examination. For most, synthesis, if it is evident, starts in youth with an attitude of fulfilling dreams by making things or expressing a rich imagination.

Our education and training systems need to consciously work to improve synthesis as well as analysis abilities. In mathematics we have analytic geometry, analytical chemistry, differential calculus—all of which are followed, if at all, by another analytical course, or in some cases by a synthesis course, as differential calculus is followed with integral calculus (I am probably one of the few people who received academic credit for integral calculus before differential calculus).

Discovery in natural systems is generally the role of the scientist steeped in analytical means, in contrast to invention of physical systems, generally the role of the professions, particularly the engineer, and some applied scientists, who use synthetic means and innovation in the economical system. The engineer's approach is more heuristic, that is a solution is found that works without having scientific proof. One can be, must be, innovative, creative, and imaginative whether in discovery or invention, but it would seem that the opportunities are greater in invention, at least in the coming century.

Creative thinking is more restrained in analytic thinking as compared to the synthetic approach where there are usually many more paths, an expanding number as one moves outward. The 'rules' for analysis tend to 'rein one in' as compared to the procedure for expanding one's horizons for synthesis. In the study of plants, one begins with the parts of the plant, the details of its components, primarily analysis in nature. When that approach is combined with synthesis, the evolution of the plant, its habitat, relationship to other plants and animals, the subject comes multifarious..

The difficulty of understanding and interpreting analysis and synthesis becomes evident as one attempts to describe analytic synthesis (Is this *a priori* synthesis?), synthetic analysis (Is this *a posterior* synthesis?), analytic analysis, and synthetic synthesis. There is a parallel with the difficulty of defining engineering science, scientific engineering (like genetic engineering), and engineered science (like engineered materials).

Plato could be classified as an early synthesist (Emerson said that Plato was a great generalizer); Aristotle, an analyst. Synthesis could go only so far, then analysis was necessary and has predominated for the last 200 years. We are now at a stage where synthesis is needed in the physical world to increase the value of analysis, or at least the direction given to the work of analysis. The philosophy may precede the physical and social aspects, as existentialism and transcendentalism are a part of synthesis.

Although synthesis and synthetic thinking were developed in early philosophical thinking, Kant through his voluminous writings, brought visibility and attention to the concepts. It wasn't until the last half of the 20th Century that the need for more emphasis on synthesis became apparent as changes in the physical world occurred, and as synthesis was recognized in the biological world as an important phenomenon, and more recently in the business world.. These events have brought synthesis to the forefront:

» Avalanche of results of analysis are available, many of which are presented as facts and data, rather than information,

» Large constructed systems and artifacts, based on synthesis of data available, built before "analysis" could be made; importance of safety and danger to people and animals and the environment; Large systems involve more synthesis.

» When analysis is done first, usually with the objective of getting data, those data do not necessarily fit the system need, leaving gaps in the data, then synthesis is used as a guide to identifying needed data.

» Increased complexity of today's manufactured or built systems; increased connectivity (or increased importance of connectivity) of different systems; not restricted to physical (artifacts) and natural (nature) but also includes human systems (social, psychological, and business)

SPIRALING

Bruner (1963) writes that early teaching should "be designed to teach these subjects with scrupulous intellectual honesty... with an emphasis upon the intuitive grasp of ideas... on the use of basic ideas, and to revisit these basic ideas repeatedly, building on them... " He emphasizes the importance of a "'spiral curriculum' that turns back on itself at higher levels." The same approach applies to the analysis-synthesis, with each drawing from and building on the other.

ACADEMIC DEPARTMENTS

Because of their single-mindedness for analysis, academic departments are often severely criticized, and rightly so. These departments in sciences are usually analysis-minded. People in other fields depend on many science departments to teach analytic methods. Departments other than science departments are needed for building inter-, multi-, and cross-disciplinary activities, are often more synthesis-minded. Disciplinary departments need to incorporate a greater respect for synthesis and reduce their attack on other departments or units that need to incorporate more synthesis in their teaching and research. Those departments that are considered more synthesis in approach must not eliminate teaching of analytic skills-meaning that education is built on both. So the approach for science disciplines should be analysis-synthesis, and the approach for the applied and professional fields should be synthesis-analysis. These approaches would be one step toward reducing the two cultures concept.

To encourage academic study in interdisciplinary fields, for which there are an increasing number, Programs (with a capital P) have often become stronger than departments from the standpoint of attracting students, granting degrees, and meeting user needs. Quality control must be a part of these needed Programs to assure a rigor appropriate to the subject. To the extent that programs are incorporated in departments, the departments are strengthened.

To eliminate analysis from educational programs would be a tragedy creating a situation far weaker than the one we have today where the analysts have taken control. A program permitting each to flourish would provide an even stronger education program than we have today.

Educators and policy people should have at least one subject in depth to experience the rigor, importance, and use of analysis. To this add an experience in breadth, with a comprehensive (synthesis) way of thinking so that analytic approaches will be within a framework of the useful or user. Some people should be encouraged to pursue strictly analytic approaches, but caution is expressed when an entire field, such as mathematics, does so. Likewise, when an entire field like environmental science does not include subjects in depth of an analytic nature–chemistry, physics, mathematics, or environmental engineering does not include courses in biology, biochemistry, and ecology, the purported program of study is misleading. Environmental studies (as a separate field) are often available with a limited analytical science base.

Most university academic departments are disciplinary in nature, and those disciplines are usually taught from an analytic approach. The appeal is to expand activities of the basic discipline by connecting to other disciplines or fields such as engineering, a nexus. In particular disciplines of engineering should incorporate related subjects and take a horizontal view as well as a vertical view. The department provides a strength, usually in analysis, needed to have a strong synthesis. Without that strength synthesis could be shallow and unproductive. In addition to departments, institutes, center, and laboratories can be initiated to develop multidisciplinary or cross-disciplinary fields, usually requiring synthesis. In universities, those additional organizations flourish if leadership and support are provided in which incentives regard the synthesis approach. The incentives for encouraging synthesis in a university and professional and technical societies are different than for analysis, and include recognizing:

1. The important role of the user in the process of producing a scholarly product;
2. That review articles are a major contribution to knowledge and literature and the development of the author;
3. The need for inter-, multi-, and cross-disciplinary activities;
4. The value of integration of knowledge and thought; and
5. The role of synthesis in identifying important work in analysis.

MEANING OF TRUTH

An interesting contrast between lawyers and scientists is represented by the meaning of truth. Two categories of truth are adversary truth and scientific truth. The lawyer's approach is argumentation and the truth sought is adversary truth. Adversary truth, as presented in debate or in court, usually includes one aspect of truth. In adversary truth arguments, uncertainties are highlighted and the uncertainties are left to the audience or court to decide. Decisions based on adversary truth are made by those not involved in the subject. People with expertise in the subject under consideration are often rejected for serving on juries considering the case. The 'truth' is decided by the jury or audience.

In the search for scientific truth people delve into the uncertainties attempting to clarify facts, data, and procedures. People repeat experiments to verify results leading to the truth. When a consensus of the community accepts, the subject is accepted as the scientific truth. At a latter date new ideas, more accurate or new instruments, and additional findings might lead to new truths.

The engineer is often in a situation that is between these two truths in which both adversary truth and scientific truth are involved, perhaps illustrating an in-between position, technological truth. Technological truth is doxa, truth beyond a shadow of doubt, and is difficult to explain to adversaries. Public hearings, expert witnesses at a trial, and liability claims in which engineers get involved incorporate aspects of both adversary and scientific truths. A malfunctioning device, an unsafe piece of equipment, an inefficient process, a polluting manufacturing process–all of these must be considered in the context of the human and social context and the political environment–not on the basis of scientific truth. The observer, participant, and user can be frustrated by the different bases for truth in our society.

RECOGNIZING THAT THE AGE OF SYNTHESIS IS UPON US

Once that synthesis is recognized as an important component in our society certain actions should be taken:

» Develop a philosophy of engineering (and other professions and some applied sciences) based on synthesis as a predominate activity;

» Augment educational programs to include synthesis;

» Organizations should identify a synthesist as well as analyst to assume a balanced, forward looking approach to identification and consideration of issues;

» Attitudes will develop to encourage people to think and act, where appropriate (more appropriately) outside the conventional (or mainstream) of thinking which we might otherwise hold onto too long. Synthesis inspires thinking ahead and encourages an attitude of considering a larger picture, a larger system;

» Learn to work together, in pairs, teams, groups, with individuals each contributing their expertise and attitudes. Ironically, more cooperative work will be involved to meet high standards and competition; and

» Postmodern and New Age concepts will need a stronger synthesis to tie into possible related events and phenomenon.

SUMMARY

Discovery (science), invention (engineers and related), and innovation (economics-technologists) are basic to the advance of our technological society and a part of the human endeavors. People contribute to and benefit from those endeavors which have become stronger and more important with the increase in age of our civilization. The search for truth was a major goal in those early years. The search for truth continues but has expanded from truth as it relates to religion to truth as it relates to natural things and systems and to truth as it relates to artificial things and systems.

The search for truth utilized several approaches: deduction and induction; tuition and intuition; syllogism; logic. These approaches were cloaked in various movements such as dogmatism, existentialism, idealism, materialism and mechanism, phenomenologism, positivism, pragmatism, rationalism, transcendentalism, and utilitarianism. These approaches and processes used to search for truth have served well in the past but need to be considered in a new Matrix, as illustrated by present trends, to more effectively meet the needs of the future. The emphasis in past years has been primarily on analysis with secondary consideration for synthesis. There has been a rapid increase over the past quarter of a century in the use of synthesis, particularly in the physical world, and the stage is set for major efforts to emerge based on synthesis. Synthesis provides a matrix (framework) for meeting human needs through technological processes, while reaching for what might be as well as what is.

Instead of basing philosophy on such concepts of deduction and induction, our thinking, teaching, and practice should be based on analysis and synthesis. Analysis and synthesis would incorporate deduction and induction, intuition, inference, logic, and whatever is needed to proceed. In thinking synthetically, one needs to develop a more global view, consider a larger system, often involve human, natural, or physical systems simultaneously. Inductive approaches do not provide that kind of matrix for thinking, but are a part of the thinking process. Rules, axioms, or postulates are needed to guide synthetic thinking. A start could be provided by some of the theorems developed for systems theory that could be generalized. And these need to be taught and discussed to better prepare the scientists and engineers for the future.

The use of THOUGHT (ideas, imagination) represented as THESES (words, designs, sketches, treatises) are steps to providing artifacts of that thought process as THINGS (artwork, vehicles, substitute items, manufactured goods, parts, systems, etc.).

The flow from THOUGHTS – to – THESES – to – THINGS driven by end use is synthesis. Preceding the resulting artifact, someone must conceive and have a vision for the end product. That conception occurs in the thought process of synthesis using deduction or induction, intuition and creativity, as needed. Synthesis is more appropriate for meeting future needs. Synthesists will set the agenda for analysts during the next century. When a stone mason who spent his days arduously chipping stone was asked what he was doing, he replied, 'building a great cathedral.' That person was a synthesist.

Part 9.
Addendum

PART 9.1
GLOSSARY RELATED TO ANALYSIS AND SYNTHESIS

apodictic – views or opinions clearly proven or demonstrated; doxa would be less certain.

a fortiori – stronger, with stronger reason, or more conclusively.

a posteriori – knowledge dependent on experience based on sights into essential relationships within the empirically given. (Runes, 1968); empirical in nature.

a priori – (statements) independent of experience; we know the truth independent of experience, or prior to experience.

analytic – or "critical, impulse treats any subject or topic by concentrating upon the part, by taking it apart in the service of clarity and precision." (Aristotle, Hume, Carnap, Russell, Moore) G. W. Leibniz (1646 – 1716) claimed that all true propositions are analytic.

an emergent – see **emergent**

animism – the theory that (a) parts are adapted (coordinated, adjusted) to other parts within a whole in order to achieve the purpose of the whole, and (b) the whole itself sets the purpose and causes the adaption. (Angeles, 1981)

aristotelian – a philosophy based on fundamental principles of formal logic, especially through the doctrine of syllogism. (Webster Third New Int. Dict., 1986)

cartesianism – a philosophy of Rene Descartes, its proclamation of mathematical certitude of logic as an ideal in metaphysical demonstration and its dualistic distinction between thought and extension of mind and matter. (Webster Third New Int. Dict., 1986)

casuistry – the art and science of making present decisions based on past decisions. (Hall, 1989)

causation, cause and effect – "A cause has traditionally been thought of as that which produces something and in terms of which that which is produced, its effect, can be explained. (Edwards, 1972) 'Cause and effect' is associated with mechanistics and materialistics.

cognitive – Study of the relationship between knowledge of fact and another feature, such as behavior.

convention – the highest level of hypothesis; above principle; above phenomena; above generalization. (Poincare, 1902)

deduction – starts with a general principle and shows its application to particular facts; when a hypothesis has been derived to fit observed facts from which various consequences can be deduced; from general to particular.

determinism – represents thought that for everything that happens "there are conditions such that, given them, nothing else could happen." (Edwards, 1972) The belief that all events have causes and occur by necessity.

dialectical – discussion between two opposed views, ultimately each person accepting part of the views of another to come to a final view that does justice to all, providing a new unity, a synthesis of views. A thesis – antithesis approach developed by Hegel. (1770 – 1831). Then Marx (1818 – 1883) transformed this from spiritual reality to materials into a dialectical materialism.

dogmatism – a view based on insufficiently examined premises; philosophy grounded on principles established by reason to the neglect of consideration of experience. (Webster Third New Int. Dict., 1986)

doxa – opinions supported by evidence and reason; knowledge in the sphere of doubt but not beyond the shadow of doubt. (Adler, 1993) Synthesis is based on doxa to a greater extent than on episteme. Also see **apodictic.**

dualism – a view that two fundamental entities can be used to represent an area of concern, such as good and evil, mind and body, and analysis and synthesis.

eclectic – selecting what appears to be the best or true from various and diverse doctrines or methods, rejecting a single and exclusive doctrine or method, in constant search for the best doctrine or method. This was the method of Aristotle. (Webster Third New Int. Dict.,1986)

emergent – the new qualitative synthesis produced by structures organized in certain patterns that cannot be predicted from the examination of the constituent parts of the whole (emergent property of water could not be predicted from the knowledge of hydrogen and oxygen). (Angeles, 1981)

empirical – relying or based on factual information, or direct sense experiences, as opposed to theoretical knowledge; experiential, observation,factual. Webster Third New Int. Dict., 1986) Empiricism emphasizes experiences and sense perceptions as the source of knowledge. John Locke was the first great empiricist in 1660, followed by George Berkeley and David Hume in the 1700s. Locke, Berkeley, Hume, Mill were leading empiricists.

entelechy – a non-material purpose or drive or force that is supposed to cause a thing to come into existence. (Martin, 1991)

episteme – knowledge with certainty without a reasonable doubt.

epistemology – "dealing with the theory of knowledge and derivation of knowledge, the scope of knowledge..." (Flew, 1979) It aims to determine the nature, basis, and extent of knowledge, and explores the relationships between knowledge and belief. Made up of correspondence, pragmatic, coherence, theories, and skepticism. (World Book, 1990)

ethics – concerns human conduct and values, in general, not to a given end. (Lacey, 1986). There are three categories of ethics: relativism (depends on culture), objectivism, and subjectivism. (World Book, 1990)

existentialism – deepest meaning of human experience as the unifying theme. J–P. Sartre, K. Jaspers, and E. Husserl. (Runes, 1968) Existential judgments are always synthetic. (Edwards, 1972) Became influential in the mid – 1900s, in which people made decisions with limited knowledge and time (as required for an engineer), in which human choice is subjective.

experimental science – Can be analytical as well as synthetic. The three towering personalities were Copernicus, Kepler, and Galileo who were champions of experimental science.

gestalt – "a unified whole that has parts that act in an organized fashion, that is greater than the sum of the parts and above the interaction of its individual and that is able to affect the behavior of its parts." (Angeles, 1981)

heuristic – providing assistance or an approach in discovering a truth or solving a problem, that is neither asserted nor evaluated as true but is assumed to be representative for specific purposes because of its previous usefulness as an investigative tool. (Angeles, 1981)

holistic – considering phenomena in terms of the functions of a whole that guides the parts; explaining the actions of parts in terms of the whole.

hypothesis - "a theory or suggestion that still has overtones of the arbitrary or speculative." (Flew, 1979)

ideas – "the distinction between abstract and concrete ideas is virtually the distinction, misleadingly put, between concepts and percepts." (Edwards, 1972)

idealism – a metaphysical theory about the nature of reality, a doctrine that reality is in some way mental. The contrast is Realism, and possibly Materialism. (Lacey, 1986)

induction – process of drawing inferences and arriving at a general principle about a whole class based on observations of a few of its members from particular fact to general principle. (Rosenberg, 1972) Francis Bacon was a strong proponent of the inductive method of investigation.

inference – "deriving conclusions from premises..." (Reese, 1980) "A rule for construction of components which says what may be inferred from one or more statement whose logical structure is specified." (Flew, 1979)

instrumentalist – one who adheres to the concept that the significance of a thing is its value as an instrument; that ideas are instruments of action and their usefulness determines truth. (Webster Third New Int. Dict., 1986) John Dewey was an instrumentalist.

intentionality – the ability of consciousness to create a mental object, refer it to reality, and direct activity to results. (Angeles, 1981)

intersubjectivity – the property of holding in reference to more than one subject, rather than purely subjectivity. (Schlick, M., from Flew, 1979); a kind of interdisciplinary.

intuition – comprehension of a subject, idea, or relationship without reasoning. Emerson was a proponent of intuition over tuition. "Tuition taught us the material world through our senses." (Bode, 1981) Intuitionism is sometimes called constructivism. (Casti, 1990)

logic – internally–related, as contrasted to externally–related; a system of formal principles of deduction or inductive reasoning.

materialism – the view that everything is made of matter, that the mind and spirit are illusionary or causally linked to matter, usually denying abstract. Hegel's dialectical materialism makes economic considerations both the cause and the proper end of human action. (Lacey, 1986)

mechanistic – a search for mechanical models, a universe of the LaPlacian type that is governed by strictly deterministic laws, in which qualities are eliminated in favor of quantifiables. (Flew, 1979) Descartes was a mechanist, cause and effect for physical phenomenon.

metaphysics – something that deals with what is beyond the physical or the experimental; relating to what is conceived as transcendental, not based on experience nor physics. (Webster Third New Int. Dict., 1986) The theories of metaphysics include materialism, idealism, and teleology. (World Book, 1990)

noetic – pertaining to the mind or thought, apprehended by reason. The science of pure thought reasoning.

ontology – "The branch of metaphysical inquiry concerned with the study of existence or being itself, considered apart from the nature of any existent object." (Flew, 1979)

organicism – represents the opposite of mechanism and represents an analogy of a living organism.

phenomenological – the study of how things appear; concept of intentionality, of consciousness toward objective, objects of experience. (Husserl, most recent, modern adherent). Study of phenomena in contrast to things in themselves, relies heavily on intuitive insights. (Seligman, 1970) Phenomenology also leads to the study of being associated with existentialism.

philosophical analysis – that one solves philosophic problems through analysis of language or concepts, of which strong advocates were B. Russell and L. Wittgenstein.

philosophy – deals with "physically" unprovable relationships, with proof based on reasoning and discussion. In its early stages, philosophy, as the study of transcendent. (Runes, 1968) Philosophy cannot, like mathematics, proceed synthetically, it must analyze and clarify. (Kant) A search for underlying causes and principles of reality and explanation of these. Until the 1700s science and philosophy were together; after the 1700s, natural philosophy (physics) and moral philosophy (ethics). The first engineering degree at West Point was in Natural Philosophy. The five branches of philosophy are: metaphysics, epistemology, logic, ethics, esthetics, and often a sixth, language. (World Book, 1990)

platonic – that ultimate reality consists of transcendent eternal universals which are the true objects of knowledge, that objects of the senses are not completely real but participate in the reality of the ideas; Plato did little for the development of empiric science. (Webster Third New Int. Dict., 1986)

positivism – thesis is that "science is the only valid knowledge and facts the only possible objects of knowledge; considers things that can be sensed; that philosophy does not possess a method different from science...opposes metaphysics and any procedures of investigation that is not reducible to scientific method..." (Edwards, 1972) Logical positivism (Carnap 1891 – 1970), a movement of the 1920s, including the thought that science and the use of logic are the source of knowledge and that metaphysics is meaningless. Comte (1798 – 1857) is considered the founder of positivism, one of the most influential movements of the 20th Century.

posteriori – see ***a posteriori***

pragmatism – "the judgment of the reality of things and ideas according to their usefulness and practical applicability." Charles Peirce (1839 – 1914), 1888; William James (1842 – 1910) (Runes, 1968) and John Dewey (1859 – 1952). "Look away from primary substances, principles and look to the end of things, the fruits, the consequences, and the actualities." That knowledge is subordinate to actions, and meaning and truth are related practice or action, developed in 1900s.

priori – see ***a priori***

rationalism – the belief or theory that the exercise of reason rather than empiricism provides the only valid basis for action or belief. (American Heritage Dictionary, 1969). Knowledge can be derived from reason itself, independent of senses. A movement of the 1600s that reason is superior to experience as a source of knowledge, used deduction, lead by Descartes (mechanist), Spinoza (axioms) and Leibniz (differential calculus).

realism – a view that emphasizes the existence or role of a thing or object as contrasted to a view in form of words or ideas.

socratic – a method of inquiry consisting of a series of questioning the object of which is to elicit a consistent expression of something supported to be implicitly known by all rational beings. (Webster Third New Int. Dict., 1986) (Also, related to development of the inductive method.)

syllogism – a deductive logical scheme or analysis of a formal argument consisting of a major premise, a minor premise, and a conclusion.

synechism – blending of inharmonious elements.

synectic – to hold together, to join, to connect apparently irrelevant elements. (Gordon, 1961)

synergy – the word that represents behavior of a whole system unpredicted by the separate behavior of any of its parts. (Fuller, 1972)

syntax – putting together.

synthetic – or "speculative, impulse operates by seeking to comprehend the whole, or by putting it all together in the service of unity and completeness." (Plato, Hegel, Whitehead, St. Thomas, Spinoza). Used to describe a statement that is neither analytic or self-contradictory. (Edwards, 1972)

synthetic philosophy – try to systematically cover all knowledge. (Spencer, 1897)

synthetistes – a group of Belgium composers (1925) to synthesize the elements of contemporary music. (Sodia, 1980)

system – a whole made up of parts, each hierarchal level having a specific identity, but without which the system would not exist. A system is identified by interactions, interconnections, and interdependence of parts as well as complexity.

tautology – means "literally, saying the same thing." (Lacey, 1987). "a logical and not a factual truth," (Flew, 1979) or any analytic truth.

teleology – "the fact or character of being directed to an end or shaped by purpose; a whole conceived as determined by final causes as with design or purpose, as opposed to a purely mechanical determinism. The world exhibits teleological order (design, adaptation), therefore it is produced by an intelligent designer; a doctrine that ends, final causes, or purposes are to be involved as principles of explanation." (Edwards, 1972) (Reese, 1980) Teleological animism means that parts within the whole are adapted to achieve the purpose of the whole, with the whole itself setting the purposes and cause adaption. (Angeles, 1981)

theism – belief in the existence of one God who is viewed as the creative source of new, etc., as opposed to atheism.

transcendentalism – a philosophy in which the principles transcend the "senses" for a fuller meaning beyond the limits of experience. Knowledge is not limited solely to that derived from experience and observation. Kant (1724 – 1804) was a transcendentalist (World Book, 1990). Ralph Waldo Emerson (1803 – 1882) was an exponent of transcendentalism. Other American transcendentalists were B. Alcott, Margaret Fuller, Theodore Parker, and Henry David Thoreau.

utilitarianism – the thought that we should always do what will produce the greatest good, as espoused by G. E. Moore (1873 – 1958) and J. S. Mill (1806 – 1873). Mill is also an empiricist and phenomenologlst.

PART 9.2
PHILOSOPHERS AND OTHERS, ANALYSISTS AND SYNTHESISTS

These are subjective classifications in which a few people, along with their major writings, are classified in one or the other category, even though they may have changed their approach during their career. Few people, if any, can are entirely in one category only.

Aquinas, St. Thomas » 225 – 1274 *(Synthesis)*
Considered by many as an analyst, a follower of Aristotle – "Summa Theologica," that considered a larger system "Posterior Analytics"

Archimedes » c.287 – 212 BC *(Synthesis)*
Greek mathematician, engineer, inventor

Aristotle » 384 – 322 BC *(Analysis)*
Deduction as a method of science; discoverer of inductive thinking; an observer rather than experimenter – "Metaphysics"; "Physics"; "Organon"

Austin, John » 1790 – 1859 *(Analysis)*
Jurisprudence analytic doctrines – "The Province of Jurisprudence Determined" (1832) "Lectures on Jurisprudence" (1863, posthumously)

Bacon, Francis » 1561 – 1626 *(Synthesis)*
Believed that scientific reasoning was by induction; forerunner of empiricist tradition – "Advancement of Learning" (1605) [use knowledge as base] "Novum Organum" (1620) [induction method]

Bacon, Roger » c.1214 – 1292 *(Synthesis)*
Pioneer in experimentation – "Opus Majus" c.1266

Boole, George » 1815 – 1864 *(Analysis)*
Deductive logic in algebra "The Mathematical Analysis of Logic" (1847) "An Investigation of the Laws of Thought" (1854)

Braun, Werhner von » 1912 – 1977*(Synthesis)*
Experiment, then analyze, in rocketry

Carnap, Rudolf » 1891 – 1970 *(Analysis)*
Classical positivist; helped found logical positivism; Interest shifted to induction. – *(Synthesis)* – Editor – "International Encyclopedia of Unified Science" "Logical Foundations of Probability" (1950)

Carnot, Nicholas Sadi » 1796 – 1832 *(Synthesis)*
Generalization that lead to Rudolph Clausis developing the concept of entropy and the second law of thermodynamics

Descartes, Rene » 1596 – 1650 *(Analysis)*
Mathematical reasoning was the paradigm for his system of knowledge; knowledge is a unified system; A rationalist; mechanistic philosophy "Discourse on Method" (1637); "The Mediations" (1641); *(Synthesis)* "Principles of Philosophy" (1644)

Dewey, John » 1859 – 1952 *(Synthesis)*
Synthesized the views of Pierce and James developing pragmatism as a theory of logical and ethical analysis; experience determines the problems to be solved – "The Quest for Certainty" (1929); "Experience and Nature" (1925)

Diderot, Denis » 1713 – 1784 *(Synthesis)*
Encyclopedic; rejected metasphysical; believed that doubt was beginning of wisdom – "Encyclopedia" (1749)

Edison, Thomas » 1847 – 1931 *(Synthesis)*
An inventor with 1093 patents; a 'cut and try' approach often used, based on experience and intuition; approach is known as Edisonian

Einstein, Albert » 1879 – 1955 *(Synthesis)*
Natural philosopher, theoretical physics; world view; his views lead to changes in concepts dealing with time and space. Modified Newtonian and Cartesian concepts – "Theory of Relativity" (1905); "The Foundation of the General Theory of Relativity" (1916)

Emerson, Ralph Waldo » 1803 – 1882 *(Synthesis)*
Transcendentalist; Imagination important attribute – "Nature" (1836); Essays; Journals

Epicurus » 341 – 270 BC *(Synthesis)*
First declared empiricist based on truths of experience Knowledge drives from sensations Diogenes has preserved three of his letters, fragments of – "On Nature"

Fuller, R. Buckminster » 1895 – 1983 *(Synthesis)*
Designer, builder – "Ideas and Integrities" (1963), Synergetics 1, 2 (1975, 1979)

Galileo Galilei (known by his first name) » 1564 – 1646 *(Synthesis)*
Built first telescope, based on work of others, and improved it until it was the best known "The Starry Messinger" (1610) "Dialogue Concerning the Two Chief World Systems" (1632)

Hegel, Georg Wilhelm F » 1770 – 1831 *(Synthesis)*
Unification of knowledge; phenomenology – "Phenomenology of the Mind" 1807 "Logic" (1812)

Hume, David » 1711 – 1776 *(Synthesis)*
Discoveries of experience, philosophic skepticism (humanism), restrict human knowledge to experience of ideas and impressions; An empiricist and naturalist; also an atomist; followed up on Francis Bacon's approaches. "A Treatise of Human Nature" (1739)

Husserl, Edmund » 1859 – 1938 *(Synthesis)*
Phenomenologist; trancendentalist – "Philosophy of Arithmetic" (1891) Several books by others based on his phenomenology – "Phenomenology and the Foundations of Science" (1980)

Kant, Immanuel » 1724 – 1804 *(Analysis)*
Transcendental dialectic; wrote widely on analytic/synthetic statements – "Critique of Pure Reason" (1781) – "Critique of Practical Reason" (1788) *(Synthesis)*

Kepler, Johannes » 1571 – 1630 *(Synthesis)*
Used intuition to develop second law of planetary force. Laid basis for Newton to discover law of gravitation. Synthesis of astronomy and physics.

Built on the data of Tycho Brahe, an analyst – "Magnus opus" (1609); "A New Astronomy Based on Causation or a Physics of the Sky" (1618)

Kettering, Charles F. » 1876 – 1958 *(Synthesis)*
Inventor; leader in engineering research

Leibniz, Gottfried W. 1646 – 1716 *(Analysis)*
Combined mathematical studies and philosophic thinking – Associated with ideas of Aristotle; Believed body works as if it had no soul, and vice–versa "New Physical Hypothesis" (1671) [Artist and scientist in one personality]

Leonardo da Vinci » 1452 – 1519 *(Synthesis)*
Artist, architect, engineer – "Treatise on Painting" (1651); many others wrote about him and his works

Locke, John » 1632 – 1704 *(Synthesis)*
Sees the fundamental unit of knowledge to be that of intuition, like Descartes; an empiricist (Flew) – "An Essay Concerning Human Understanding" (1690); "Two Treatises of Government" (1690)

Medawar, Peter B. » 1915 – 1987 *(Analysis)*
Believed that inductive thinking not scientific – "Induction and Intuition in Scientific Thought" (1969); "The Limits of Science" (1984)

Michelangelo » 1475 – 1564 *(Synthesis)*
One of world's greatest artists

Mill, John Stuart » 1806 – 1873 *(Analysis)*
Induction involves a search for causes; an empiricist – "A System of Logic" 2 vol. (1843); "On Liberty" (1859)

Moore, George E. » 1873 – 1958 *(Analysis)*
Anti-thesis, thesis, etc; viewpoint is one of essentially rejecting the skeptical conclusions often drawn from empiricism – "Principia Ethica" (1903)

Newton, Isaac » 1642 – 1727 *(Synthesis)*
Claimed that his method was empirical and inductive, in contrast to Descartes "Principia" (1687); – "Mathematical Principles of Natural Philosophy"

Plato » 427 – 347 BC *(Synthesis)*
Interested in moral philosophy and despised natural philosophy (Teacher of Aristotle); a rationalist (Flew); A synthesizer greater than Socrates (Emerson, p. 316) – " Republic"; " Dialogues of Plato"

Quine, Willard van Orman » 1908 – *(Synthesis)*
There must be nonanalytic truths – most influential American empiricists – "Two Dogmas of Empiricism" (1951); "On What There Is" (1953)

Russell, Bertrand » 1872 – 1970 *(Analysis)*
System of mathematical logic; logical atomism – "Principia Mathematica" 3 vol. (1910, 1912, 1913) with Whitehead; "Introduction to Mathematical Philosophy" (1919)

Sartre, Jean – Paul » 1905 – 1980 *(Synthesis)*
Existentialist type of synthesis – "The Psychology of Imagination" (1940), "Existentialism and Humanism" (1946)

Socrates » c.470 – 399 BC *(Synthesis)*
Ethics, virtue, inquiry and instruction. Search for what is implicitly known – "Dialogues"

Spencer, Herbert » 1820 – 1903 *(Synthesis)*
Synthetic philosophy. Wrote – "A System of Synthetic Philosophy." 1897 Try to systematically cover all knowledge. Began as a Civil Engineer – "First Principles" (1862); "Principles of Biology" (1864 – 1867)

Teilhard de Chardin, Pierre » 1881 – 1955 *(Synthesis)*
A strong visualizer – "The Phenomenon of Man" (1959)

Tillich, Paul » 1886 – 1965 *(Synthesis)*
An amalgamation of existentialism, psychology, and neo–scholastic ontology – "Systematic Theology" 3 vol. (1951 – 1963)

Whitehead, Alfred North » 1861 – 1947 *(Synthesis)*
Considered the world as a comprehensive metaphysical system to include a scientific cosmology "Principia Mathematica" (1910 – 1913) with Russell, "Science and the Modern World" (1925)

Wittgenstein, Ludwig » 1889 – 1951 *(Synthesis)*
First studied engineering; theoretical limit of analysis – "Tractatus" (1922) [a metaphysical work]; "Philosophical Investigations" (1953 tr.)

PART 9.3
PHILOSOPHIC IDEAS AND THEIR PROPONENTS/ADHERENTS

Analysts [tend to distrust those who construct theories of 'reality as a whole']

Aristotle (384 – 322 BC) [deduction as a method of science]
Boole, George (1815 – 1864) [deductive logic in algebra]
Leibniz, G. W. (1646 – 1716) [all true propositions are analytic]
Russell, Bertrand (1872 – 1970) [great logician; philosophic analyst]
Moore, George E. (1873 – 1958) [analysis is not words but concepts]

Wittgenstein, Ludwig (1889 – 1951) [philosophic analyst; studied

Atomists [subdivide into the smallest part; also associated with materialists]

Democritus (460 – 370 BC). [none of his works survived in original form.]
Epicurus (341 – 270 BC) [took views of Democritus]
Galileo Galilei (1564 – 1642) [found atomism compatible with his work]
Boyle, Robert (1627 – 1691) [chemical properties of atoms]
Dalton, John (1766 – 1844) [chemical combinations and properties]
Avogadro, Amedo (1776 – 1856) [related volume and number of atoms]
Russell, Bertrand (1872 – 1970) [logical atomism]

Deductionists [it is impossible to assert premises and to deny the conclusions without thereby contradicting oneself]

Descartes, Rene (1596 – 1650) [mathematical reasoning is the paradigm for new system of knowledge – Cartesian]
Newton, Isaac (1642 – 1727) [claimed that his method was both empirical and intuitive]
Jaskowski, S. (1906 – 1965) [natural deduction]
Gentzen, G. (1909 – 1945) [natural deduction]

Dualists [dualism asserts that substances are either material or mental; Cartesian philosophies are dualistic]

Zoroaster (c.660 BC);
Plato (428 – 348 BC);
Descartes, Rene (1596 – 1650)
Kant, Immanuel (1724 – 1804);
Hegel, Georg W. F. (1770 – 1830)

Empiricists [knowledge of factual matters based on experience]

Bacon, Francis (1561 – 1626) [forerunner of the empiricist tradition]
Locke, John (1632 – 1704) [first great empiricist]
Berkeley, George (1685 – 1753) [his doctrine was that there is no material substance and that things are a collection of ideas or sensations which exist in the world]
Hume, David (1711 – 1776) [developed his arguments from experience]
Diderot, Denis (1713 – 1784) [followed Locke; opposed metaphysics]
Mill, John S. (1806 – 1873) [induction thinking involves search for cause, using empirical thinking]
James, William (1842 – 1910) [known for radical empiricism]

Experimentalists [can be analytic as well as synthetic] [experimentalism is name given by J. S. Mill to his form of empiricism] (Reese, 1980)

Copernicus, Nicolaus (1473 – 1543) [his work supported the idea that the earth was spherical and that the earth moved around the sun]
Galileo Galilei (1564 – 1642) [ability to develop thought experiments, like Einstein]
Kepler, Johannes (1571 – 1630) [also an empiricist]

Existentialists [an approach or alternative challenging essentialism, rationalism, empiricism]

Kierkegaard, Soren (1813 – 1855) [doctrine of subjective truth]
Nietzsche, Friedrich (1844 – 1900) [challenged Hegelian philosophy; emphasized individual and importance of subjectivity]
Husserl, Edmund (1859 – 1938) [he strongly influenced movement]
Teilhard de Chardin, Pierre (1881 – 1955) [and transcendental speculations]
Jaspers, Karl (1883 – 1969) [one of founders of existentialism]
Tillich, Paul (1886 – 1965) [he drew mainly from Heidegger]
Barth, Karl (1886 – 1968) ["Belief cannot argue with unbelief ..."]
Marcel, Gabriel (1889 – 1973) [communication between God and man]
Heidegger, Martin (1889 – 1976) [modern approach]
Sartre, Jean–Paul (1905 – 1980) [leading exponent of atheistic existentialism]

Inductionists [moved from various bits of evidence to a general principle]

Bacon, Francis (1561 – 1626) [inductive logic][forerunner of empiricist tradition]
Whewell, William (1794 – 1886) [there can not be a logic of induction parallel to a logic of deduction; wrote two books on induction]
Mill, John S. (1806 – 1873) [used logic method of induction]
Popper, Karl R. (1902 –) [explains science and the scientific method; rejected inductive method in empirical sciences; hypotheses are deductively proven by trying to discover exceptions]

Instrumentalists [opposed to most realist theories of science]

Dewey, John (1859 – 1952) [stresses instruments and means over ends]
Nagel, Ernest (1901 –) [teleological explanations are equivalent to nonteleological ones in biology and psychology]

Intuitionists [contrasted with empiricism and rationalism as source of knowedge]

Brouwer, L. E. J. (1881 – 1966) [mathematical entities constructed from intuition]
Heyting, A. (1898 – 1980) [intuition used for number theory]

Materialists [goes back to Epicurus] [Also see Atomists]

Hobbes, Thomas (1588 – 1679) [founding father of modern metaphysical materialism]
Diderot, Denis (1713 – 1784) [endowed matter with sensitivity]
Priestly, Joseph (1733 – 1804) [discovered oxygen; theology is compatible to religion]
Hegel, Georg W. F. (1770 – 1831) [dialectical materialism]
Marx, Karl (1818 – 1883) [dialectical materialist method]
Engels, Friedrich (1820 – 1895) [thesis – antithesis – synthesis pattern; attributed to Fichte, John G.]

Mechanists [based on quantifiables, purposes not to be included in scientific practice]

Hobbes, Thomas (1588 – 1679) [also see materialist]
Descartes, Rene (1596 – 1650) [cause and effect]
Newton, Isaac (1642 – 1727) [Newtonian mechanics]
Leibniz, Gottfried W. (1646 – 1716) [each substance is independent of other substance]
Hume, David (1711 – 1776) [also see empiricists] [also a materialist]

Phenomenologists [consider actual or possible 'perceptual' experiences]

Hegel, Georg W. F. (1770 – 1831) [phenomenology of the mind]
Mill, John S. (1806 – 1873) [social living corresponds to fixed laws]
Mach, Ernst (1836 – 1916) [philosophy is radically empiricist – logical]
Bretano, Franz (1838 – 1916) [started phenomenology movement]
Husserl, Edward (1859 – 1938) [father of phenomenology][experience is not limited to sensual apprehensions]
Heidegger, Martin (1884 – 1976) [adopted Husserl's method]

Positivists (Logical Positivists) [Vienna Circle, most important influence on the 20th Century]

D'Alembert, Jean Baptist (1717 – 1783)
Comte, Auguste (1798 – 1857) [considered father of positivism; only knowledge worthwhile was knowledge that could be used]
Spencer, Herbert (1820 – 1903) [philosophy distinguished by its generality]
Russell, Bertrand (1872 – 1970) [construction out of known entities for inference to unknown ones]

Neurath, Otto (1882 – 1945) [physicalism, the view that all empirical statements can in principle be formulated as statements, not necessarily experience]
Schlick, Moritz (1882 – 1936) [analysis of experiment, preserve intersubjectivity]
Wittgenstein, Ludwig (1889 – 1951) [originally an engineer who developed 'picture' of thing described by language and as seen]
Carnap, Rudolf (1891 – 1970) [helped to found logical positivism in the 1920s]
Popper, Karl (1902 –) [developed reasoning based on probability]
Ayer, A. J. (1910 –) [provided a link between several philosophies]

Pragmatists [consider that effects might have practical bearings]

Peirce, Charles (1839 – 1914) [developed the doctrine of pragmatism]
James, William (1842 – 1910) [one of first pragmatists]
Dewey, John (1859 – 1952) [leading exponent of pragmatism]
Schiller, F. C. S. (1864 – 1937) [truth and reality are manmade]
Lewis, C. I. (1883 – 1964) [pragmatic theory of the *a priori* problems]
Nagel, Ernest (1901 –) [subscribed to naturalism and instrumentalism]
Quine, W. V. (1908 –) [prefers a physicalistic system]

Rationalists [appeal to reason as a source of knowledge; Plato, early view was of rational persuasion; rational emphasizes the *a priori* , but not intuition]

Descartes, Rene (1596 – 1650) [mechanist; favored deductive method]
Spinoza, Baruch (1632 – 1677) [axioms; rationalist philosopher, deductive proof]
Liebniz, Godfrey (1646 – 1716) [differential calculus]
Voltaire, Francis M. (1694 – 1778) [a major figure in the Enlightenment]
Kant, Immanuel (1724 – 1804) [as well as Transcendental Idealism]
Medawar, P. G. (1915 – 1987) [believed that inductive thinking not scientific]

Reductionists or Reductivists [generally a doctrine that claims to reduce the apparently more sophisticated and complex to the less so; the attitude of the mind that one believes that he/she keeps in touch with reality if the outlook is narrowed to obtain specific facts, preferably facts that can be put in numbers; human behavior can be determined on basis of lower animals]

Pavlov, Ivan (1849 – 1936) [with dogs];
Skinner, B. F. (1904 – 1990) [with rats and pigeons];
Lorenz, K. (1903 – 1989) [with geese]
Relates to Phenomenalists, other Empiricists, and Positivists

Synthesists

Plato (427? – 347 BC) [one of the greatest generalizers – Emerson]
St. Thomas Aquinas (1225 – 1274) [his general dialectic begins from induction and proceeds deductively either to a posteriori for discovery (inventions) or *a priori* for assessment; Thomism developed from his philosophy]
Hobbes, Thomas (1588 – 1679) [his philosophical ideas of a metaphysical speculation have a second order statement that says something general about a whole class of first order statements]
Hegel, George W. F. (1770 – 1831) [an objective realist][thought culminates in art, religion, and philosophy] [approach characterized by use of dialectic – thesis, antithesis, synthesis; world develops by the interplay of opposites, duality]
Whitehead, Alfred North (1861 – 1947) [nature is nothing more than can be observed through the senses; he applies concepts that originate in study of organisms to interpretation of physics to psychology, etc]

Transcendentalists [dependence on experience or reason]

Kant, Immanuel (1724 – 1804) [transcendentalist idealist]
Emerson, Ralph Waldo (1803 – 1882) [wrote a manifesto of transcendentalism called Nature; self – reliance]
Thoreau, Henry (1817 – 1862) [follower of Emerson]

Utlilitariansts [greatest good for the greatest number]

Bentham, Jeremy (1748 – 1832) [founder of English utilitarian philosophy]
Mill, John S. (1806 – 1873) [wrote book on utilitarianism in 1863]
Spencer, Herbert (1820 – 1903) [a civil engineer, then was an utilitarian philosopher]
Moore, George E. (1873 – 1958) [common experience yields a great deal of knowledge]

Sources: Edwards (1972); Flew (1984); Lacey (1986); Reese (1980)

PART 9.4
EXAMPLES OF PEOPLE WITHIN LAST CENTURY WITH SYNTHESIS APPROACH AND TECHNOLOGICAL ACCOMPLISHMENTS

Some of the outstanding synthetic technological accomplishments during the past century.

Armour Research Institute, Chicago, IL
Commercial development of optical fibers, 1960.

Arrhenius, Svante (1859 – 1927)
Explained the greenhouse effect, 1908

Asimov, Isaac (1920 – 1993)
Synthesized information from several fields and presented imaginative views for future.

Backus, John
Development of FORTRAN, the first widely used general-purpose computer language, 1957

Baekeland, Leo H. (1863 – 1944)
Invented bakelite, 1909.

Bardeen, John; Brattain, Walter; and Shockley, William
Developed transistor, 1947.

Bell, Alexander G. (1847 – 1922)
Designed and developed the telephone.

Benz, Carl (1844 – 1929)
Developed early automobile, 1893.

Bernard, Christiaan (1922 –)
Performed heart transplant, 1967.

Bjerknes, Vilhelm (1862 – 1951)
Developed meteorology base and weather forecasting.

Bohr, Nils (1885 – 1962)(analysis)
Discovered structure of atom.

Braun, Wernher von (1912 – 1977)
Developed rocketry.

Broglie, Louis–Victor de (1875 – 1960)
Developed ideas on light wave.

Bronowski, Jacob (1908 – 1974)
Art and science as "twins" in expression of human imagination.

Carlson, Chester F. (1906 – 1968)
Copying device, 1938.

Carothers, Wallace H. (1896 – 1937)
Invented nylon, first true manmade fiber, 1931.

Chadwick, Sir James (1891 – 1974)
Discovered neutron, 1932.

Cockcroft, Sir John Douglas (1897 – 1967)
Invented particle accelerator.

Cockerell, Sir Christopher (1910 –)
Hovercraft (travel on cushion of air), 1959.

Cormack, A. (1924 –) and Hounsfield G. (1919 –)
Developed computerized axial tomography (CATScan), 1973.

Cousteau, Jacques – Yves (1910 –)
Developed aqualung, 1943.

Cray, Seymour B. (1925 –)
Built first supercomputer, 1976.

Crick, Francis (1916 –)
Co–discoverer of double helix of DNA, 1953.
Also see James Watson.

Crookes, Sir William (1832 – 1919)
Design of cathode ray tube, c. 1887.

Curie, Marie (1867 – 1934)
Discovery of radium and work in radioactivity.

Diesel, Rudolf (1858 – 1913)
Invented diesel combustion cycle and engine, 1892.

Dunlap, John Boyd (1840 – 1921)
Developed rubber pneumatic tire, 1887.

Eastman, George (1854 – 1932)
Developed photography, flexible roll film.

Edison, Thomas Alva (1847 – 1931)
Invented numerous devices, received 1093 patents.

Ehrlich, Paul (1854 – 1915)
Developed first synthetic drug called arsphenamine, effective for syphillis, c.1910.

Einstein, Albert (1879 – 1955)
Formulated relativity; intuitive synthesist.

Einthoven, Willem (1860 – 1927)
Invented electrocardiograph, 1911.

Farnsworth, Philo, and Zworykin, Vladimir
Invented television, c. 1935.

Fermi, Enrico (1901 – 1954)
Developed first nuclear reactor, 1942.

Feynman, Richard (1918 – 1988)
Innovative comprehensive thinker.

Fleming, Sir Alexander (1881 – 1955)
Discovered penicillin (result of laboratory accident), 1928.

Ford, Henry (1863 – 1947)
Innovations of assembly line production; first motorcar in 1896 (built on work of many before him: Eli Whitney, George Eastman).

Freud, Sigmund (1865 – 1939)
Related the unconscious and conscious; a physician and a scientist; a mechanist and a determinist.

Fry, Arthur L.
Developed Post–it note™ in 1974 at 3M Co.

Fuller, R. Buckminster (1895 – 1983)
Synthesist of ideas and things.

Goddard, Robert H. (1882 – 1945)
Invented first liquid fuel rocket, 1926.

Haber, Fritz (1868 – 1934)
Developed chemical synthesis of ammonia for fertilizer, 1908.

Hahn, Otto (1879 – 1968)
Work leading to the first nuclear reactor.

Joliot–Curie, Frederic (1900 – 1958)
Joliot–Curie, Irene (1897 – 1956)
Discovered means of creating artificial radioactivity.

Judson, Whitcomb L. (1846 – 1909)
Invented device for slide fastner, later called zipper, 1891.

Kapany, Narinder S.
Developed optical fibers, 1955.

Kettering, Charles Franklin (1876 – 1958)
Inventions such as cash register, automotive electrical equipment, quick drying paint, antiknock combustion additives.

Kilby, Jack S. (1923 –) and Robert N. Noyce (1927 – 1991).
Invented integrated circuits, 1959; commercially, 1964; Draper Award in 1990.

Land, Edwin (1909 –)
Invented polarized camera; instant developing of photographs, 1948.

Landsteiner, Karl (1868 – 1943)
Discovered how to make blood transfusions, discovered ABO blood groupings and compatibility of certain bloods, c.1900.

Laue, Max von (1879 – 1960)
Developed X–ray crystallography, c. 1912.

Lister, Joseph (1827 – 1912)
Inventive surgeon, pioneered in antiseptics.

Lumiere, Auguste (1862 – 1954)
Lumiere, Louis (1864 – 1948)
Invention of motion picture camera and projector, 1895.

Marconi, Guglielma (1874 – 1937)
Synthesized all the ideas on wireless communications and added his own ideas to make the first working radio telegraph, c.1901.

Mendeleyev, Dmitry (1834 – 1907)
Synthesized knowledge of 63 known chemical elements to form the periodic table, 1869.

deMestral, Georges
Inventor of velcro, 1948.

Michelson, A. A. (1852 – 1931)
Contributed to knowledge of light theory.

Neumann, John von (1903 – 1957)
Prepared foundation for computers.

Nipkow, Paul (1860 – 1940)
Invented scanning disc, basis of TV, 1884.

Ohain, Hans von (1911 –)
Developed jet engine, 1939. Draper Award in 1992.
Also see Sir Frank Whittle.

Parsons, Sir Charles P. (1854 – 1931)
Invented steam turbine, c.1884.

Pavlov, Ivan P. (1849 – 1936)
Discovered conditioned reflex.

Perkin, Sir William Henry (1838 – 1907)
Developed synthetic dye, aniline purple, for dyeing cloth, 1856.

Planck, Max (1858 – 1947)
Developed modern quantum theory, 1900.

Roentgen, Wilhelm (1845 – 1923)
Discovered X–ray, 1903.

Rubic, Erno
Invented cube puzzle game, 1979.

Rutherford, Sir Ernest (1871 – 1937)
Described radioactive decay, published, 1904.

Sabine, Wallace Clement (1868 – 1919)
Developed empirical approach to auditorium acoustics, c. 1900.

Salk, Jonas (1914 –)
Developed first vaccine effective against polio, published results in 1953.

Sikorsky, Igor E. (1889 – 1972)
Invented helicopter, first flew in 1939.

Steinmitz, Charles (1865 – 1923)
Invented devices to use alternating current electricity.

Telsa, Nikola (1856 – 1943)
Built first alternating current induction motor, 1893.

Thomson, Sir J. J. (1856 – 1940)
Discovered the electron; developed knowledge of electromagnetic spectrum.

Townes, Charles H. (1915 –)
Invented MASER, microwave generator used in long distance communications, 1953.

Van Der Waals, Johannes D. (1837 – 1923)
Formulated equations on behavior of gases.

Wankel, Felix (1902 –)
Invented rotary piston engine, 1957.

Watson, James (1928 –)
Co–discoverer of double helix of DNA, 1953.
Also see Francis Crick.

Watson–Watt, Sir Robert (1892 – 1973)
Applied theory of radar, 1935.

Whittle, Sir Frank (1907 –)
Developed turbojet engine, 1930s. Also see Hans von Ohain.

Wright, Orville (1871 – 1948)
Wright, Wilbur (1867 – 1912)
Invented first practical powered airplane capable of controlled flight, 1903.

Zeppelin, Ferdinand Graf von (1838 – 1917)
Rigid, lighter than air device, first flight, July 2, 1900.

Numerous important works of engineers are carried out by a team or group of people. Appropriate credit may not go to the critical individuals who contributed creatively to the project. Project leaders or chief engineers are often recognized for the leadership provided, who in turn should provide appropriate credit and recognition to the members of the team. The IBM 360 computer system, under the leadership of Eric Block; the space flight to the moon, under the leadership of Werhner von Braun; and the development of the interstate highway system in the United States, led by Francis (Frank) Turner, are a few examples.

REFERENCES:

Asimov, Isaac. 1972.
Biographical Encyclopedia of Science and Technology. Avon Book, New York, NY 805 p. Q141.A74

Devine, Elizabeth, Held, Michael, Vinson, James, and Walsh, George. 1983.
Thinkers of the Twentieth Century: A Biographical and Critical Dictionary. Macmillan, London. 643 p. B804.T456

Feldman, Anthony and Ford, Peter. 1979.
Scientists and Inventors.
Facts on File, New York, NY 336 p. T39.F3

Giscard d'Estaing, Valerie – Anne. 1985.
The World Almanac Book of Inventions. World Almanac Publications, New York, NY. 361 p. T15.G57

Parkinson, Claire L. 1985.
Breakthroughs: A Chronology of Great Achievements in Science and Mathematics, 1200 – 1930. G. K. Hall, Boston, MA 576 p. Q125.P327

Stone, Judith. 1988. Velcro: the final frontier. Discover 9(5):82 – 84. May.

PART 9.5 REPRESENTATIVE STATEMENTS OF RELATIONSHIPS OF VARIABLES

Analogy – A statement that embodies a proportion or ratio, as in mathematics, or correspondence or agreement in value or meaning. [Napier's Analogy]

Aphorism – A concise statement of a principle, truth, or opinion; a maxim; an adage.

Assertion – A belief or an idea; subject to proof or verification. [Mayer Assertion (energy)]

Axiom, Maxim – Self–evident or accept as self–evident; Newton's first two laws originated as axioms; axiom of choice in statistics; a basic proposition in a formal system presented without proof. Wolff described axioms as indemonstratable, theoretical, and universal, thus belonging to the synthetic. See postulate. [Archimede's Axiom; Peano's Axiom]

Canon – General rule or generally accepted basic principle. [Canon of Ethics; Canon of Rules on Induction (Mill); Lloyd Morgan's Canon]

Concept – An idea expressed in words. [Copernican Concept; Gamow's Concept; Newton's Concept]

Conjecture – To infer from incomplete information or evidence. [Goldbach Conjecture; Ramanujan Conjecture]

Convention – A belief widely accepted by long usage. [Convention Truth]

Corollary – A proposition that follows with little proof from another already proven; a subsidiary result that can be derived from a theorem.

Criterion – A principle against which a proposition can be treated or tested. [Lawson's Criterion (for fusion); Raleigh's Criterion; Routh's Criterion]

Dilemma – A situation faced with choices between unsatisfactory solutions. [Doctor's Dilemma; Prisoner's Dilemma]

Doctrine, Dogma – Idea put forth as true, not necessarily supported by evidence. [Doctrine of Signature]

Effect – A response to a stimulus. [Doppler Effect; Greenhouse Effect; Joule–Thomson Effect; Mossbauer Effect; Raman Effect]

Equation – A statement of equality between two expressions. [Arrhenius Equation (rate reaction); Drake Equation; Hamilton Equation (position of a body); Lange – Hicks Equation; Van der Pol Equation]

Formula – A general principle stated in mathematical terminology. [Cauchy Formula; Clausius Formula; Leibnitz Formula; Planck's Formula]

Function – A mathematical relationship. [Bessel Function; Dirac Delta Function; Gibb's Function; Goedel's Function; Green's Function]

Hypothesis – A statement of conjecture subject to proof; hypothesis is that inertial forces arise from the gravitational interaction between the accelerated body and mass of the distant stars. [Avogadro's Hypthesis; Continuum Hypothesis; Gaia Hypthesis]

Inequality – A statement of inequality between expressions. [Camp – Meidell Inequality (called a law by E. B. Wilson)]

Law – A general principle that represents all cases in an applicable situation. [DeMorgan's Laws; Distributive Law (of logic); First, Second, Third Laws of Thermodynamics; Ohm's Law]

Model – Tentative representation for testing a phenomenon (structure, equation, relationship, etc.). [Charney's Model (of Navier–Stokes Equation); Dyson's Model; Linear Model; Richardson's Model; Sommerfield Model (metal)]

Paradox – A true statement that appears to be contrary to other statements or opinions. [Berry Paradox (complexity); D'Alembert Paradox; Grue–Bleen Paradox; Russell's Paradox; Zeno's Paradox]

Phenomenon – An observable fact or event. [El Nino Phenomenon; Magnetoelastic Phenomenon; Relaxation Phenomenon]

Postulate – An observation or projection of a relationship based on insight or intuition that becomes a hypothesis if proved to be true; often used synonomously with axiom; but Wolff described postulates as demonstratable, practical and particular, thus belonging to the analytic. [Koch's Postulates (bacteriology)]

Principle – A basic truth, law, or assumption. [Complementarity Principle; Copernican Principle; Hamilton Principle; Heisenberg Uncertainty Principle;

Pauli Exclusion Principle]

Problem – A question raised for discussion and inquiry. [Abel's Problem; Dido's Problem; Halting Problem; Hilbert's Problem; Ozone Problem; Three Body Problem]

Proof – Expressions dealing with mathematical symbolic logic. [Goedel's Proof]

Proposition – A statement. A deductive demonstration of a mathematical statement.

Relation(ship) – A thought advanced as true. [Abbe Relation; Gibb's Relation; Newland's Relationship (became known as the Law of Octaves); Onsager Relation; Preditor–Prey Relation; Rankine–Hugoniot Relation]

Rule – Standard procedures for solving a class of problems. [Geiger – Nutal Rule; Gibb's Phase Rule; Kramer's Rule; L'Hopital Rule; Rule of signs (algebraic); Simpson's Rule]

Scenario – an outline of an expectedactivity. [Eigen Scenario (biology); Fox's Scenario (biology); Gilbert Scenario (biology)]

Scientific Method – An approach that combines deduction and induction.

Statement – As applied to computer programming, a component of a program that gives instructions to the computer. [GOTO Statement]

Supposition – The property of a term whereby it stands for something

Tautology – A compound proposition that is true regardless of values assigned, such as "A or not A"; cannot be false. [Laws of Thought]

Tenet – A thought or idea advanced as true. [Central Tenet of Human Sociobiology]

Theorem – An idea demonstratably true or proven proposition. [Chebyshev's Theorem (called the Law of Large Numbers); Goedel's Theorem (a well formed formula in a given logic system); Maclaurin's Theorem; Pythagoras Theorem]

Theory – System of circumstances leading to explanation of a phenomenon. [Einstein's Relativity Theory; Quantum Theory; Theory of Gravitation]

Thesis – A proposition; subject of a composition; forming ideas into expressible form to be communicated to others, usually written but can be spoken. [Church's Thesis; Turing-Church Thesis]

NOTE

These definitions are not clear cut. Usage often determines the name given. For example, Newland's Relationship became a law; Delta function is not a function; Hamilton Principle is not a principle; many laws are really axioms, such as the Laws of Conservation of Energy.

Representative examples of the relationships are given. The author has collected over 1500 named laws from the scientific and technology literature.

PART 9.6
QUOTATIONS RELATING TO SYNTHESIS

Ardigo, Roberto (1828 – 1920).

"The difference between self and things is nothing more than the difference between 'autosynthesis' and a 'heterosynthesis' of neutral elements."

Source: *Reese (1980).*

Bloom, Allan. 1987.

"Our desire for conflict reduction accounts for the great popularity of the word 'dialectic' – in our sense, the Marxist sense – for, beginning in opposites it ends in synthesis, all charms and temptations united in harmony... Socratic dialectic takes place in speech and, although drawn forward by the search for synthesis, always culminates in doubt... Man is not just a problem–solving being ...but a problem–recognizing and-accepting being."

Source: *The Closing of the American Mind. p. 228.*

Bordogna, Joseph. 1990.

"A good job has been done during the later half of this century in educating our students in analysis – in independently picking things apart and understanding the pieces. Our task now is to emphasize the need to synthesize, to integrate, to put things together as a team – while continuing to do so in a strong academic setting."

Source: *Issues in Science and Technology 7(1):20.*

Bordogna, Joseph. 1992.

"Whereas scientists might be thought of as lords of focused depth, an engineer's duty is to be societies master integrator. Both aptitudes are equally creative endeavors and both require scholarship of equal value and of the highest standards."

Source: *Bent 83(3):26 – 27.*

Branscomb, L. M. 1986.

"...the use of advanced engineering technology to further scientific objectives...the theoretical experiments of the scientists were hard to distinguish from the prototype simulations of the engineer. (Instruments)... attained a degree of control and a complexity beyond the reach of science 20 years ago."

Source: *The American Scientist 74(6):650 – 658.*

Chekov, Anton (1860 – 1904). 1888.

"An artist observes, selects, guesses and synthesizes."

Source: *Bartlett (1992). p. 582.*

Coleridge, Samuel Taylor (1772 – 1834). 1817.

"He [the poet] diffuses a tone and spirit of unity, that blends, and (as it were) fuses, each into each, by that synthetic and magical power... imagination."

Source: *Barlett (1992). p. 384.*

Descartes, R. (1596 – 1650).

"Analysis is more *a priori.* Synthesis is *a posteriori* to demonstrate conclusion clearly, and employs a long series of definitions, postulates, theorems – see Part 9.5 for a summary of these – so that if challenger denies one of the conclusion, it can be shown at once that it (conclusion) is contained in what went before. However this method is not as satisfying as the method of analysis – nor is the learning mine as engaged or as interested in synthesis; analysis and the generations of the old could keep to themselves (at least they made it appear as a kind of mystery with synthesis), whereas with analysis, parts are learned."

Source: *Selected Philosophical Writings (1984).*

Dostoevski, F. M. (1821 – 1881).

"Before the problem of the creative writer, analysis need lay down arms." [Let synthesis reign]

Source: *Bartlett (1992). p. 569.*

Drexler, K. Eric. 1986.

"Assemblers will help engineers synthesize things; their relatives, disassemblers, will help scientists and engineers analyze things."

Source: *Engines of Creation, p. 19.*

Dutton, D. and Krausz, M. 1981.

"The more familiar each of the previously unrelated structures are, the more striking the new synthesis and the more obvious it seems in the driver's mirror of hindsight."

Source: *The Concepts of Creativity in Science and Art, p. 2.*

Emerson, Ralph Waldo (1803 – 1882).

"Succession, division, parts, particles – this is the condition, this is the tragedy of man. All things cohere and unite. Man studies the parts, strives to tear the part from the connection, to magnify it, and make it whole."

Source: *Journals V. 1838, pp. 83 – 84.*

Emerson, Ralph Waldo (1803 – 1882).

" ...the analytic process is cold and bereiving (sic) and shall I say it? somewhat mean, as spying... The poet sees wholes and avoids analysis... I think that philosophy is still crude and elementary. It will one day be taught by poets."

Source: *Natural History of Intellect, p. 13.*

Emerson, Ralph Waldo (1803 – 1882).

"The wonderful synthesis so familiar in nature... Every great artist has been such by synthesis... The synthesis which makes the character of his (Plato) mind appears in all his talents."

Source: *Bode (1981). pp. 304 – 305.*

Engles, Friedrich (1820 – 1895).

"Without analysis, no synthesis."

Source: *Mackal (1991).*

Forrester, Jay W. 1967.

"They (engineers) act as the interfaces between compartments of human behavior. They must resynthesize the fragments caused by specialization of others."

Source: *Engineering Education and Practice in the Year 2000.*

Fuller, R. Buckminster (1895 – 1983). 1963.

"It is the function of engineering to provide society with reliable predictions as to the behavior characteristics of complex design predicated on competent expertise. Engineering... consolidates the net gains of science and design in the industrial complex. Gains are design intuited synergies."

Source: *Ideas and Integrities. p. 191.*

Fuller, R. Buckminster (1895 – 1983).
(synthesis) "We will be omni–integrated synergetics, viewing the part in relation to the whole."
Source: *Devine, et al. (1983). p. 192.*

Galbraith, J. K. 1983.
"The leaders of organizations are largely organizational creations and are synthetic personalities. Divorced from the organization, the synthetic personality dissolves."
Source: *Galbraith (1983) pp. 42 – 43.*

Gell–Mann, Murray. 1984.
"New subjects, highly interdisciplinary in traditional terms, are emerging and represent in many cases the frontier of research. These interdisciplinary subjects do not link together the whole of one traditional discipline with another, particular subfields are joined together to make a new subject. The pattern is a varied one and constantly changing… It could herald a new synthesis of knowledge based in some general way on the notion of complexity."
Source: *Pagels, Heinz R. (1988) p. 36.*

Glashow, Sheldon. 1988.
"Eastern cultures place considerable emphasis on the unity of nature. In Hindu prayer "OM!" signifies the oneness of Krishna and Vishnu, and the unity of earth, sky and sun."
Source: *Interactions. p. 300.*

Glashow, Sheldon. 1988.
"Superstring is the name of the game, an unlikely synthesis of some of the most bizarre theoretical notions ever put forward."
Source: *Interactions. p. 330.*

Goethe, Johann W. (1749 – 1832).
" …whoever (the deity) wants to find you in infinite space and time, must first divide (analyze) and then combine (synthesize)."

Goethe, Johann W. (1749 – 1832).
The artist may well be advised to keep his work to himself till he is completed, because no one can really help him,… but the scientist is wiser not to withhold a single finding or a single conjecture from publicity."
Source: *Bartlett (1992). p. 350 from 'Essay on Experimentation.'*

Hamerton, P. G. 1864.
"Synthesists find continual pleasure in observing the relations of things,

but from their largeness of range they constantly miss minute truth, nor do they ever see anything so vividly as the analysts see that which they have analyzed."

Source: *Fine Arts Quarterly. p. 238.*

Hegel, O. W. F. (1770 – 1830).

" ...the Hegel system is avowedly an attempt to unify opposites – spirit and nature, universal and particular, ideal and real – and to be a synthesis in which all the parties and contradictory philosophies of his predecessors are alike contained and transcended. It is both idealism and realism at one."

Source: *Encyclopaedia Britannica (1987) p. 20:553.*

Hoberman, Chuck. 1992.

"In developing ideas like mine (in creative design), it makes the most sense to start on the fantasy end of things and work toward the reality end."

Source: *Discover (1992) p. 77.*

Hoffmann, Roald. 1991.

"The synthesis of molecules puts the chemists very close to the arts and engineering. We make the objects that we study and appreciate... Synthesis is a building process... many of the most distinguished synthetic studies have entailed a balance between two different research philosophies, one embodying the ideal of a deductive analysis based on known methodology and current theory, and the other emphasizing innovation and even speculation."

Source: *American Scientist 79(1):11, 14. Jan. – Feb. 1991.*

Holton, Gerald. 1972.

"The human mind proceeds in two steps: first analysis, which produces highly specialized work, and then synthesis."

Source: *Winkler, K. (1987) p. A-1.*

Huxley, Aldous

...in response to the fact that he was receiving letters from readers troubled by their inability to correlate new discoveries in virtually any branch of knowledge... "What is needed," he said, "is greater synthesis." He said that his grandfather Thomas H. Huxley suggested the creation of a university devoted simply to do that.

Source: *Julian Behrstock. 1993.*
Letter to the International Herald Tribune, November, p. 7

Huxley, Thomas H. 1877.

"The discovery of the composition of water was indeed made originally by synthetical not analytical means."

Source: *Physiography. p. 111.*

Kant, Immanuel (1724-1804)

" ...synthesis, the generating of unities out of manifolds [a set of its elements]"

Kash, Don E. 1989.

"The new reality rests on the capability to carry out continuous synthetic innovation... Synthesis... is as much an artistic enterprise as it is a scientific one... Synthetic innovation requires experimental learning and trial and error as well as scientific knowledge and information... Synthesis relies on knowing how to do more than we understand how to do."

Source: *Perpetual Innovation. pp. 38-39.*

Kash, Don E. 1991.

"I am one of those people who believes we have made a fundamental, historical transition from an industrial society to a synthetic society... Manufacturing in this new synthetic society is quit a different thing from what it was in the industrial society. Success rests on innovation, getting there first or early, rather than on efficiency... Competitive advantage in the synthetic society requires you to get there first or in parallel with your competitors, with higher quality, superior performance and products and sometimes lower cost products... "

Source: *NCAR 44, Southwest Research Institute. p. 30.*

Kellogg, C. Gary. 1987.

"We have become a society of great analyzers, but we are declining in synthesis... If there is a failure or even a perceived failure, we are quick to analyze the causes and summarily depose the designer or synthesizer... a profession cannot continue for long unless leadership is developed that views creation as more important than criticism... "

Source: *Engineering Times 9(1):4.*

Lodge, Rupert C. 1968.

"Analysis culminates in a number of points, all distinct and unconnected. What becomes of synthesis, the principle of construction which, starting with such points, builds them into a connected whole? That the imagination is constructive and synthetic is obvious. But is not imagination, in its synthesis, often groping after something objective, some connection of which reason would approve, something universal and necessary?"

Source: *The Great Thinkers p. 252.*

"Synthesis starts with small units, generally the products of a preceding analysis, and puts them together into larger unitary wholes. It may do no more than put together what has previously been taken apart. But it is quite usual for thinking, experimenting with the elements before it, to find that it can construct new unitary wholes. Sometimes supplementation is employed; more frequently, judicious omission is all that is required."

Source: *The Great Thinkers. p. 196.*

Maxwell, James C. (1831-1879).

"One of the graduate courses that I took dealt with classical electromagnetic theory and presented the great synthesis of James Clark Maxwell... (who) unified the theories of electricity and magnetism in 1865."

Source: *Glashow, S. 1988. p. 40.*

Mayr, E., and Provine, W. B. 1980.

"A parallel development in evolutionary biology resulted in a synthesis of population and Mendelian genetics with Darwinian selection theory. The so-called evolutionary synthesis attempted to account for the origin and inheritance of variations as pertaining to the nature of species and the role of selection."

Source: *The Evolutionary Synthesis.*

Medawar, P. B 1967.

"Stewart (Dugald) said that it is by reasoning synthetically from the hypothesis, and comparing the deductions with observations and experiment... Steward's own analysis shows he is here using 'synthetically' in the sense of 'deductively.'"

Source: *The Art of the Soluble. pp. 150-151.*

"The scientific method is not deduction; scientific laws are inductive in nature, you start with a small idea and build up... induction in the wider sense that Mill gave it is a myth...induction is a myth."

Source: *The Art of the Soluble. pp. 131-133.*

Medawar, P. B. 1990.

"The schoolbook notion of a separate inductive scientific method, the one we all learned from John Stuart Mill, he (Medawar) termed a fraud." Thus, he says that the synthetic approach is not a scientific approach.

Source: *The Threat and the Glory on Science and Scientists. Book Review, p. 3, Washington Post, July 15, 1990.*

National Research Council. 1991.

"The object of design processes is to synthesize solutions, that is, to imagine separate ideas and information into a unified whole. This process of adding and integrating information and knowledge about the design is done almost continuously... Synthesis at every stage involves generating alternative solutions to the problems at hand, analyzing and integrating the information derived into the design."

Source: *Improving Engineering Design. pp. 56-57.*

National Science Board. 1988.

"Engineering is different from science. Synthesis is the aim of engineering, and the defining activity in engineering is design. Engineering employs science – both science from other fields and 'engineering science' generated internally – as means to the end of producing something new... Engineering integrates analysis with synthesis while dealing with the constraints of budgets and deadlines; concerns such as health, safety, and environmental impact; and the political and social setting generally... NSF's efforts in engineering education... should focus on reintroducing and reemphasizing synthesis as the engineer's characteristic task... integrating analysis and synthesis."

Source: *The Role of the NSF in Economic Competitiveness. p. iii.*

Ortega y Gasset, Jose (1883-1955). 1944.

" ...We need to develop a special type of talent, for the special function of synthesizing... The need to create sound synthesis and systemation of knowledge... will call out a kind of scientific genius which hitherto has existed only as an aberration: the genius for integration. Of necessity this means specialization, as all creative effort inevitably does; but this time the [person} will be specializing in the construction of the whole."

Source: *Mission of the University. pp. 40, 78, 80.*

Plato (427BC-347BC).

Emerson calls Plato "this exhausting generalizer". Plato... "finding himself still capable of a larger synthesis (more than Socrates) beyond all example, then or since."

Source: *Emerson. 1971. p. 295.*

Plato (427BC-347BC)

"You can't conceive the many without the one."

Source: *Republic VIII, p. 166. Bartlett (1992). p. 76.*

Simon, Herbert A. 1969.

" ...We speak of engineering as concerned with 'synthesis,' while science is concerned with "analysis." Synthetic or artificial objects – and more specifically, prospective artificial objects having desired properties – are the central objective of engineering activity and skill."

Source: *The Science of the Artificial. p. 5.*

"In both science and engineering, the study of "systems" is an increasingly popular activity. Its popularity is more a response to a pressing need for synthesizing and analyzing complexity than it is to any large development of a body of knowledge and techniques for dealing with complexity."

Source: *The Science of the Artificial. p. 118.*

Simon, Herbert A. 1991.

"I used to think that engineering education should be more science and mathematics... The major differences between science and the professions (engineering, medicine, etc) are largely the distinctions between analysis and synthesis."

Source: *Models of My Life. pp. 257-258.*

Spencer, Herbert (1820-1908). 1860.

"Spencer saw philosophy as a synthesis of the fundamental principles of the special sciences, a sort of scientific summa to replace the theological systems of the Middle Ages. He thought of unification in terms of development... of the biological species. Spencer's attempt to synthesize the sciences showed a sublime audacity that has not been repeated because the intellectual specialization he welcomed and predicted increased even beyond his expectation."

Source: *Synthetic Philosophy (1860).*
[As written in Encyclopaedia Britannica 11:84:1a. 1987]

Stewart, Dugald (1753-1828).

"It is by reasoning synthetically from the hypothesis, and comparing the deductions with observation and experiment..."

Source: *Medawar, P. B. 1967. p. 150.*

Tauber, Alfred. 1991.

"Within a short space of twelve years (1936-1947) the disagreements [between geneticists and naturalists-systematists] were almost suddenly shoved away and a seemingly new evolution was synthesized from the valid components of the previously feuding theories. Huxley (1942) referred to this episode in history as the 'evolutionary synthesis.' "

Source: *BioScience 41(6):417-418.*

Teilhard de Chardin, Pierre (1883-1936). 1959.

"...modern thought is at last getting acclimatized once more to the idea of the creative value of synthesis in the evolutionary sense. There is definitely more in the molecule than in the atom, more in a cell than a molecule... "

Source: *The Phenomenon of Man. pp. 267-268.*

"Man represents... the most synthesized state under which the stuff of the earth is available to us."

Source: *The Phenomenon of Man. p. 281.*

Thompson, John. 1982.

"My intent in writing [this book] is to synthesize general patterns in the evolution of interaction and in the condition likely to favor coevolution among interacting species. The emphasis throughout is on patterns that transcend taxonomic boundaries – that is, on patterns that derive more from the mode of interaction than from the biology of a particular taxonomic group."

Source: *Interaction and Coevolution. p. v.*

Weinberg, A. M. 1966.

"In the university the specialist and analyst is king... Since these [societal] problems are not generated within any single discipline, their resolution is not to be found within a single discipline... In society the non-specialist and synthesizer is king."

Source: *Keenan, B. R. 1966, pp. 162-163.*

White, Morton G. (1917-). 1955.

"...the twentieth century has witnessed a great preoccupation with analysis as opposed to the large, synthetic, system-building of some other periods in the history of philosophy."

Source: *The Age of Analysis. p. iv.*

Whitehead, Alfred N. (1861-1947). 1925.

"The actual occasion (alpha) synthesizes in itself every eternal object [A, say]; and, in so doing, it includes the complete determinate relatedness of A to every other eternal object, or set of objects. This synthesis is a limitation of realization but not of content. Each relationship preserves its inherent self-identity. But grades of entry into this synthesis are inherent in each actual occasion, such as alpha."

Source: *Science and the Modern World. p. 146.*

Whitehead, Alfred N. (1861-1947).

"A judgment is a synthetic feeling, embracing two subordinate feelings in one unity or feeling."

Source: *Mays, Wolfe (1977). p. 119.*

Wilson, E. Bright. 1972.

"Synthesis puts parts back together; this may be the only situation capable of experimental or observational test. It is important that the correct parts be assembled and the effects of their interaction be properly taken into account."

Source: *Brown, Stanley (1972) v. 2; pp. 54-66.*

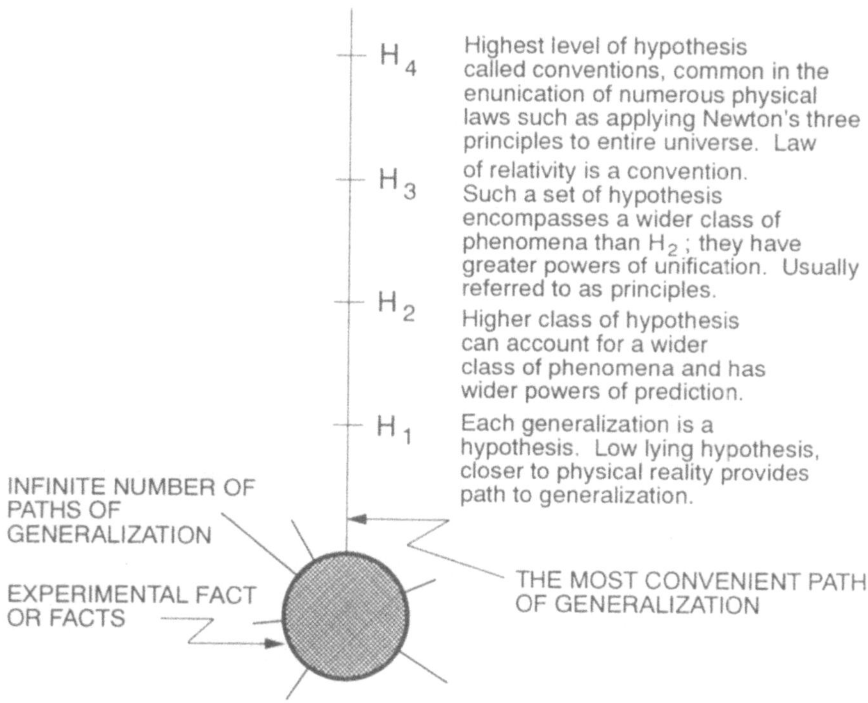

Image 9.1 Poincares Epistemology (Based on Poincare, 1902; in Miller, 1986)

Part 10.
References

PART 10.1
BOOKS

Adler, Mortimer. 1952.
The Great Ideas-A Syntopican.
Encyclopedia Britannica, University of Chicago, Il 2 vol. AC1.G75

Adler, Mortimer. 1977.
Reforming Education.
Macmillan Publishing Co., New York, NY 362 p. LA212.A37

Adler, Mortimer. 1993.
The Four Dimensions of Philosophy. Mamillan Publishing Co., New York, NY 273 p. B53.A29

Alexander, Christopher. 1964.
Notes on the Synthesis of Form.
Harvard Unversity Press, Cambridge, MA 216 p. NK1505.A4

Allen, Garland E. 1975. 1979.
Life Science in the Twentieth Century.
Cambridge University Press, New York, NY 257 p. QH305.A44

Allen, Thomas J. 1977.
Managing the Flow of Technology.
MIT Press, Cambridge, MA 320 p. T174.3.A4

Angeles, Peter A. 1981.
Dictionary of Philosophy.
Barnes & Noble Books, New York, NY. 326 p. BH41.A53

Arieti, Silvano. 1976.
Creativity: The Magic Synthesis.
Basic Books, New York, NY 448 p. BF408.A64

Ashby, Eric. 1974.
Adapting Universities to a Technological Society.
Jossey-Bass Publishers, San Francisco, CA 158 p. LB2341.A743

Asimov, I. 1989.
Asimov on Science.
Windsor Publishing Co., New York, NY 477 p. Q126.8A85

Baher, H. 1984.
Synthesis of Electrical Networks.
Wiley, New York, NY 285 p.TK454.2B.33

Bahm, Archie J., editor. 1977.
Interdependence: An Interdisciplinary Study.
World Books, Albuquerque, NM 132 p. AZ361.I57

Bartlett, J. 1992.
Familiar Quotetions.
Little, Brown and Co., Boston, MA 1405 p. PN6031.B3

Begiebing, Robert J. 1989.
Toward a New Synthesis: John Fowles, John Gardner, Norman Mailer.
UMI Research Press, Ann Arbor, MI 152 p.PS 379.B414 PS 379.B364

Bell, Daniel. 1973.
The Coming of the Post-Industrial Society.
Basic Books, New York, NY 507 p. HN17.5.B3

Berlinski, David. 1976.
On Systems Analysis.
The MIT Press, Cambridge, MA 188 p. QA402.B4

Bertalanffy, Ludwig von. 1968.
General Systems Theory: Foundations, Development, Applications.
Braziller, New York, NY 289 p.Q295.B4

Berthold-Bond, Daniel. 1989.
Hegel's Grand Synthesis: A Study of Being, Thought, and History.
State University of New York Press, Albany, NY 233 p. B2948.B46

Bloom, Allan. 1987.
The Closing of the American Mind.
Simon and Schuster, New York, NY 392 p. E169.1.B653

Bloom, Benjamin S., Krathwohl, David R., and Masia, Bertram. 1956. 1965.
Taxonomy of Educational Objectives: Handbook I: Cognitive Domain.
David McKay Co., New York, NY 2 vol. LB41.T3 LB17.T3

Bode, Carl, editor. 1981.
The Portable Emerson.
Viking Penguin Books, New York and London 671 p. PS1603.B6

Boorstin, Daniel J. 1978.
The Republic of Technology.
Harper & Row, New York, NY 105 p.T14.5B66

Bowen, Catherine Drinker. 1966.
Miracle at Philadelphia.
Little, Brown and Co., Boston, MA 346 p.JK146.B75

Boyer, Ernest L. 1990.
Scholarship Reconsidered: Priorities of the Professoriate.
Carnegie Foundation for the Advancement of Teaching.
Washington, DC 147 p. LA227.3B71

Brown, M. T. 1990.
Working Ethics.
Jossy-Bass Publishers, San Francisco, CA 219 p. HF5387.B7

Brown, Stanley B.,editor-in-chief. 1972.
The Realm of Science.
Touchstone Publishing Co., Louisville, KY 21 vol. Q111.R38

Brumbaugh, Jesse Franklin. 1917.
Legal Reasoning and Briefing.
The Bobbs-Merrill Co., Indianapolis, IN 775 p. K212.B7

Bruner, Jerome. 1963.
The Process of Education.
Vintage Books, New York, NY 97 p. LB885.B78

Buckley, N. 1968.
Modern Systems Research for the Behavioural Scientist.
Aldine Press, Chicago, IL 525 p. H83.B84

Bunge, Mario A. 1985.
The Philosophy of Science and Technology.
Kluwer Academic, Boston, MA Two volumes. Q175.B86

Bush, Vannevar. 1967.
Science is Not Enough.
William Morrow & Co., New York, NY 192 p.Q171.B95

Calvino, Italo. 1988.
Six Memos for the Next Millennium.
Harvard University Press, Cambridge, MA 124 p. PQ4809.A45A23

Carnap, Rudolf. 1952.
The Continuum of Inductive Methods.
University of Chicago Press, Chicago, IL 92 p. BC135.C312

Carnap, Rudolf. 1958.
Introduction to Symbolic Logic and Its Implications.
Dover Publications, New York, NY 241 p. BC135.C3142

Carpa, Fritjof. 1975.
The Tao of Physics.
Random House, New York, NY 346 p. QC6.C277

Cassidy, H. G. 1962.
The Sciences and the Arts--a New Alliance.
Harper & Brothers, New York, NY 182 p. Q175.C432

Casti, John L. 1979.
Connectivity, Complexity, and Catastrophe in Large-Scale Systems.
John Wiley & Sons, New York, NY 203 p. QA402.C3

Casti, John L. 1989.
Paradigms Lost.
William Morrow and Co., Inc., New York, NY 565 p. Q173.C35

Casti, John L. 1990.
Searching for Certainty.
William Morrow and Co., Inc., Nw York, NY 496 p. Q175.C434

Cauer, Wilhelm. 1958.
Synthesis of Linear Communications Networks.
McGraw-Hill, New York, NY 886 p.QC641.C323

Chang, Sheldon S. L. 1961.
Synthesis of Optimum Control Systems.
McGraw-Hill, New York, NY 381 p.TJ216.C45

Churchman, C. West. 1968.
The Systems Approach.
Delacorte, New York, NY 243 p. HD20.5C47

Churchman, C. West. 1971.
Design of Inquiring Systems: Basic Concepts of Systems and Organizations.
Basic Books, New York, NY 288 p. Q295.C46

Clavelin, Maurice. 1974.
The Natural Philosophy of Galileo.
The MIT Press, Cambridge, MA 498 p.B785.G24C513

Colley, Ann C. 1990.
The Search for Synthesis in Literature and Art.
The University of Georgia Press, Athens, GA 175 p.T385.S45

Collins, F. Howard. 1889. (Edited and consolidated)
An Epitome of the Synthetic Philosophy.
See Spencer, Herbert. 1889.

Connolly, Terry. 1983.
Scientists, Engineers, and Organizations.
Brooks/Cole Engineering Division, Monterey, CA 388 p. Q147.C65

Corey, E. J. and Cheng, Xue-Min. 1989.
The Logic of Chemical Synthesis.
J. Wiley & Sons, New York, NY 436 p. QD262.C577

Covey, Stephen R. 1989.
The Seven Habits of Highly Effective People.
Simon & Schuster, New York, NY 358 p. BF637.S8C68

Davies, Paul. 1992.
The Mind of God.
Touchstone, Simon & Schuster, New York, NY 254 P.BL240.2D29

Davis, Philip J., and Park, David. 1987.
No Way: The Nature of the Impossible.
W. H. Freeman and Co., New York, NY 325 p.Q173.N7

deCamp, L. Sprague. 1974.
The Ancient Engineers.
Ballantine, Random House, New York, NY 450 p. TA16.D4

Derry, Thomas K., and Williams, Trevor I. 1961.
A Short History of Technology.
Oxford University Press, New York, NY 782 p. T15D.4

Descartes, Rene. 1951.
Philosophical Essays: Discourse on Method, Meditations.
Bobbs-Merrill Educational Publishing, Indianapolis, IN 236 p. B1837.L3

Descartes, Rene. (translated by J. Cottingham, et al). 1984.
The Philosophical Writings of Descartes, Selections.
Cambridge University Press, Cambridge, England 2 vol.: I,418 p.; II, 428 p. B1837.1985

Devine, Elizabeth, et al. 1983.
Thinkers of the Twentieth Century.
Macmillan, London and Gale Research, Detroit, MI 643 p. CT120.T45

Dias, R. W. M. 1979.
A Bibliography of Jurisprudence.
Butterworths, London, England 149 p. K201.D5

Dobzhansky, T. G. 1974.
Studies on the Philosophy of Biology: Reduction and Related Problems.
Macmillan, London 390 p. QH331.S86

Drexler, K. Eric. 1986.
Engines of Creation.
Anchor Press, Doubleday, New York, NY 298 p. T47.D74

Durant, Will. 1961.
The Story of Philosophy.
Simon and Schuster, New YOrk, NY 412 p.B72.D8

Durant, Will and Ariel. 1967.
Rousseau and Revolution.
Simon & Schuster, New York, NY 1091 p.D286.D951

Dutton, D. and Krausz, M. 1981.
The Concept of Creativity in Science and Art.
Martinus Nijhoff Publishers, The Hague, The Netherlands 212 p. BH301.C84C66

Edwards, Paul, editor-in-chief. 1972.
The Encyclopedia of Philosophy.
Macmillan, New York, NY 8 vol. B41.E5

Eldridge, Niles. 1985.
Unfinished Synthesis.
Oxford University Press, Oxford, England. 237 p. QH366.2E53

Ellul, Jacques. 1964.
The Technological Society.
Vintage, Random House, NY 449 p.T14E51

Emerson, Ralph Waldo. 1893.
Natural History of the Intellect.
Houghton, Mifflin and Co., Boston, MA 353 p. PS1619.A.1

Emerson, Ralph Waldo. 1971.
Collected Works.
Belknap Press, Cambridge, MA 4 vol. PS1600.F71

Emerson, Ralph Waldo. 1981. (Edited by Carl Bode)
The Portable Emerson.
Penguin Books, New York, NY 688 p. PS1603.B6

Encyclopedia Britannica. 15th edition. 1974. 1987. 1993.
Encyclopedia Britannica, Inc., Chicago & London 30 vol.AG5.B8

Encyclopedia Britannica. 1992 (2nd ed.).
Great Books of the Western World.
Encyclopedia Britannica, Chicago & London 61 vol.AC.G72

Erdman, Arthur G., and Sandor, George N. 1990.
Mechanism Design: Analysis and Synthesis.
Prentice-Hall, Englewood Cliffs, NJ 2 vol. TJ230.E67

Feigl, Herbert, and Maxwell, Grover, editors. 1962.
Minnesota Studies in the Philosophy of Science. Vol III, includes, The analytic and the synthetic, pp 358-397.
Univ. of Minnesota Press, Minneapolis, MN 628 p. Q175.M64 v.3

Feldman, Anthony, and Ford, Peter. 1979.
Scientists and Inventors.
Facts on File, New York, NY 336 p. T39.F3

Ferguson, Eugene C. 1992.
Engineering and the Mind's Eye.
MIT Press, Cambridge, MA 241 p.TA145.F37

Finocchiaro, Maurice A. 1980.
Galileo and the Art of Reasoning.
D. Reidel Pub. Co., Dordrecht, The Netherlands 478 p.B785.G24F56

Flew, Antony. 1984.
A Dictionary of Philosophy.
Macmillan and Co., New York, NY 380 p.B41.D52

Florman, Samuel C. 1976.
The Existential Pleasures of Engineering.
St. Martin's Press, New York, NY 160 p.T14.F56

Foster, David. 1985.
The Philosophical Scientists.
Dorset Press, New York, NY 179 p.BT102.F67

Friedmann, W. 1967.
Legal Theory.
Columbia University Press, New York, NY 607 p. K246.F74

Frolov, Ivan T. 1986.
Man, Science, Humanism: A New Synthesis.
Prometheus Books, Buffalo, NY 341 p.BJ57.F7613

Frost, S. E. 1989 (Reprint of 1962 edition).
Basic Teachings of the Great Philosophrs.
Anchor Books, Doubleday, New York, NY 301 p. BD21.F73

Fukuoka, Masanobu. 1978.
The One-Straw Revolution.
Rodale Press, Emmaus, PA 181 p.S604F841

Fuller, R. Buckminster. 1962.
Education Automation.
Anchor Doubleday & Co., Garden City, NJ 85 p. LB885.F842

Fuller, R. Buckminster. 1963.
Ideas and Integrities.
Collier Books, New York, NY 318 p. T40.F86A3

Fuller, R. Buckminster. 1972.
Intuition.
Doubleday, Garden City, NY 190 p. PS3511.U6617

Fuller, R. Buckminster. 1976.
And It Came to Pass — Not to Stay.
Macmillan, New York, NY 157 p. PS3511.U663

Fuller, R. Buckminster. 1983.
In Thinkers of the Twentieth Century, by Elizabeth Divine.
Gale Research, Detroit, MI 643 p. CT120.T45

Galbraith, John K. 1983.
The Anatomy of Power.
Houghton Mifflin, Boston, MA 206 p. HM271.G27

Giedion, Sigfried. 1948.
Mechanization Takes Command.
Oxford University Press, New York, NY 743 p. T19.G54

Giscard d'Estaing, Valerie-Anne. 1985.
World Book of Inventions.
World Almanac Publications, New York, NY 361 p.T15.G57

Glashow, Sheldon, with Ben Bova. 1988.
Interactions: A Journey Through the Mind of a Particle Physicist and the Matter of the Mind.
Warner Books, New York, NY 347 p. QC778.G53

Goldberg, Philip. 1983.
The Intuitive Edge.
St. Martins Press, New York, NY 241 p. BD181.064

Goodfield, June G. 1981.
An Imagined World.
Harper & Row, New York, NY 240 p. QR182.5G66

Gordon, W. J. J. 1961.
Synectics, the Development of Creative Capacity.
Harper, New York, NY 180 p. BF408.G64

Gropius, Walter, 1935.
The New Architecture and the Bauhaus.
Faber and Faber, London, England 112 p NA680.G7

Guillemin, Ernst A. 1957.
Synthesis of Passive Networks: Theory and Methods Appropriate to the Realization and Approximation Problems.
Wiley, New York, NY 741 p. TK3226.G84

Hall, Arthur D. III. 1989.
Metasystems Methodology: A New Synthesis and Unification.
Pergamon Press, Oxford, England 518 p. TA168.H315

Hall, Stephen S. 1987.
Invisible Frontiers: The Race to Synthesize a Human Gene.
Atlantic Monthly Press, New York, NY 534 p. TP248.6H3

Hamerton, Philip G.(1834-1894). 1871.
Thoughts About Art.
Roberts Brothers, Boston, MA 385 p. N7445.H3

Hartshorne, Charles. 1970. 1983.
Creative Synthesis and Philosophic Method.
University Press of America, Washington, DC 337 p. B945.H335

Haykin, Simon S. 1969.
Synthesis of RC Active Filter Networks.
Wiley, London and New York 168 p. TK7872.F5H38

Hayles, N. Katherine. 1984.
The Cosmic Web: Scientific Field Models and Literary Strategies in the Twentieth Century.
Cornell University Press, Ithaca, NY 209 p. PN771.H36

Hayles, N. Katherine. 1989.
Chaos Bound: Orderly Disorder.
Cornell University Press, Ithaca, NY 309 p. PN771.H35

Hazelrigg, George A. 1993.
Systems Engineering. (prepublication copy)
Institute for Advanced Engineering, Seoul, Korea 545 p.

Heidegger, Martin. 1977.
The Question Concerning Technology.
Harper Torchbooks, New York, NY 182 p. B3279H48Q47

Hofstadter, Douglas R. 1980.
Godel, Escher, Bach.
Vintage Books, New York, NY 777 p.QA9.8 H63

Hofstadter, Douglas R. 1986.
Metamagical Themas.
Bantam Books, New York, NY 852 p.Q335.H63

Hofstadter, Douglas R., and Dennitt, D. C. 1981.
The Mind's I.
Bantam Books, New York, NY 501 p.B29.H62

Holdsworth, Ross H. 1962.
A Synthesis of Evolutionary Theory.
Prentice-Hall, Englewood, NJ 387 p. QH366.R778

Holton, Gerald. 1972.
The Twentieth Century Science.
W. W. Norton & Co., New York, NY 527 p. Q125.T84

Holton, Gerald. 1988.
Thematic Origins of Scientific Thought.
Harvard University Press, Cambridge, MA 499 p. Q125.H722

Hook, Sydney, editor. 1950.
John Dewey: Philosopher of Science and Freedom.
Dial Press, New York, NY 383 p. B945.D44H473

Horowitz, Isaac M. 1963.
Synthesis of Feedback Systems.
Academic Press, New York, NY 726 p.TJ216.H6

Husserl, Edmund C. (1859-1938). 1980.
Phenomenology and the Foundations of Science.
Martinus Nijhoff Pub., The Hague and Kluwer, Boston, MA 130 p.
B3279.H93P4713

Hutcheon, Linda. 1988.
A Poetics of Postmodernism: History, Theory, Fiction.
Routledge, New York, NY 268 p. PN3503.H83

Huxley, Julian (1887-1975). 1942.
Evolution, the Modern Synthesis.
Harper & Brothers, New York, NY 645 p .QH366.H85

Huxley, Thomas H. (1825-1895). 1877. 1878.
Physiography, An Introduction to the Study of Nature.
Macmillan, London 423 p. GB54.H9

Ihde, Don. 1978.
Technics and Praxis-A Philosophy of Technology.
D. Reidel, Dordrecht, The Netherlands 151 p.Q174.B67

James, William. 1936.
The Varieties of Religious Experiences.
The Modern Library, New York, NY 526 p. BR110.J3

James, William. 1967.
Essays in Radical Empiricism and a Pluralistic Universe.
P. Smith, Gloucester, MA 232 p. B945.J23E7

Jencks, Charles (editor). 1980.
Post-Modern Classicism: The New Synthesis.
Architectural Design, London, England 142 p NA682.P67P69

Johnson, Steven D. 1984.
Synthesis of Digital Designs from Recursion.
The MIT Press, Cambridge, MA 207 p. TK7868.D5J84

Juenger, Friedrich Georg. 1956.
The Failure of Technology. (The Perfection of Technology, the original title.)
Henry Regnery Co., Chicago, IL 204 p.T14.J814

Kant, Immanuel (1724-1804). 1949. (Lewis W. Beck, editor)
Critique of Practical Reason. (First edition, 1788)
University of Chicago Press, Chicago, IL 369 p.B2758.B353

Kant, Immanuel (1724-1804). 1952.
Critique of Pure Reason. (First edition, 1781)
(Volume of the Great Books of the Western World).
Encyclopedia Britannica, Inc., Chicago, IL 613 p. AC1.G786

Kash, Don E. 1989.
Perpetual Innovation.
Basic Books, New York, NY 269 p. HC110.T4K37

Keenan, B. R. (ed.). 1966.
Science and the University.
Columbia University Press, New York, NY 207 p. Q183.5.P9K4

Kerruish, Valerie. 1991.
Jurisprudence as Ideology.
Routledge, London; and New York, NY 221 p.K370.K47

Kidder, Tracy. 1981.
The Soul of a New Machine.
Atlantic-Little, Brown and Co., Boston, MA 293 p. TK7885.4K53

Kiernan, Thomas P. 1965.
Who's Who in the History of Philosophy.
Philosophical Library, New York, NY 185 p. B41.K5

Kirk-Othmer. 1984.
Encyclopedia of Chemical Technology.
John Wiley & Sons, New York, NY 24 vol. TP9.K54

Klein, Julie Thompson. 1990.
Interdisciplinarity: History, Theory, and Practice.
Wayne State University Press, Detroit, MI 331 p. BD255.K54

Koen, Billy V. 1985.
Definition of An Engineering Method.
American Society for Engineering Education, Washington, DC 79 p. T60.6.K63

Kuhn, Thomas S. 1962.
The Structure of Scientific Revolutions.
University of Chicago Press, Chicago, IL 172 p.Q175.K95

Lacey, A. R. 1986.
A Dictionary of Philosophy.
Routledge & Keagon Paul, London 266 p.B41.L32

LaFollette, Marcel Chotkowski, editor. 1982.
Quality in Science.
The MIT Press, Cambridge, MA 169 p.Q172.5.S34Q34

Lampugnani – see Magnago-Lampugnani

Le Corbusier – see Stanislaus von Moos. 1979.

Le Corbusier. (1887-1965). 1946. 1986.
Towards a New Architecture.
Architectural Press, London and Praeger, New York, NY 289 p.NA2520.J431986

Lee, Irving J. 1954.
Customs and Crises in Communication.
Harper, NY 334 p.PN4181.L42

Lodge, Rupert C. 1945.
Philosophy of Business.
University of Chicago Press, Chicago, IL 432 p.HF5353.L6

Lodge, Rupert C. 1947.
Philosophy of Education.
Harper, New York, NY 350 p. LB875.L6

Lodge, Rupert C. 1968.
The Great Thinkers.
Kennikat Press, Port Washington, NY 310 p. B72.L8

McGraw-Hill. 1987.
McGraw-Hill Encyclopedia of Science and Technology.
McGraw-Hill Book Co., New York, NY 20 vol. Q121.M3

Mackay, Alan L. 1991.
A Dictionary of Scientific Quotations.
Adam Hilger, Philadelphia, PA: Bristol, England 297 p Q173.H36

Magill, Frank N., and McGreal, I. P. 1961.
World Philosophy.
Salem Press, Englewood Cliffs, NJ 8 vol. B21.M3

Magnago-Lampugnani, Vittorio; Bergdoll, Barry; Hatje, Gerd; Petje, Wolfgand.1986.
Encyclopedia of 20th Century Architecture.
N. N. Abrams, New York, NY 4 vol. NA680.H3913

Maheu, Rene, et al. 1967 – see UNESCO.

Margenau, Henry, editor. 1972.
Integrative Principles of Modern Thought.
Gordon and Breach, New York, NY 522 p. BD218.M37

Margolis, Joseph. 1987.
Science Without Unity.
Basil Blackwell, Oxford, England 470 p.Q175.M36

Martin, Robert M. 1991.
The Philosophy Dictionary.
Broadview Press, Peterborough, Ontario, Canada 255 p. B41.M37

Mayr, E., and Provine, W. B. 1980.
The Evolutionary Synthesis: Perspectives on the Unification of Biology.
Harvard University Press, Cambridge, MA 487 p. QH366.2E87

Mays, Wolfe. 1977.
Whitehead's Philosophy.
Martinus Nijhoff, The Hague, The Netherlands 144 p. Q175.M414

Medawar, Peter B. 1967.
The Art of the Soluble.
Meuthuen & Co., London 160 p. Q175.M42

Medawar, Peter B. 1969.
Induction and Intuition in Scientific Thought.
American Philosophical Society, Philadelphia, PA AC1.J39
69 p. Q11.P612

Medawar, Peter B. 1984.
The Limits of Science.
Harper & Row, New York, NY 108 p. Q175.M4

Medawar, Peter B. (1915-1987)1990. (Edited by David Pyke)
The Threat and the Glory: Reflections on Science and Scientists.
HarperCollins, New York, NY 291 p. Q175.M435

Messadie, Gerald. 1991.
Great Modern Inventions.
W. R. Chambers Publ., New York, NY 276 p. T18.M4713

Meyers, R. A., editor. 1987.
Encyclopedia of Physical Science and Technology.
Academic Press, Orlando, FL 18 vol. Q123.E497

Michener, James A. 1988.
The Journey.
Randam House, New York, NY 244 p. PS3524.119J68

Mill, John Stuart. 1884 (Eighth edition).
A System of Logic, Ratiocinative and Inductive, Being a Connected View of the Principles of Evidence and the Methods of Scientific Investigation.
Harper & Bro., New York, NY 659 p. BC91.M6

Miller, Arthur I. 1986.
Imagery in Scientific Thought.
The MIT Press, Cambridge, MA 355 p. Q175.M628

Mohr, Hans. 1947.
Lectures on the Structures and Significance of Science.
Springer Verlag, New York, NY 227 p. Q171M687

Moos, Stanislaus von. 1979.
Le Corbusier, Elements of a Synthesis.
The MIT Press, Cambridge, MA 379 p.NA1053.J4M613

Moranian, Thomas. 1963.
The Research and Development Engineer as Manager.
Holt, Rinehart, & Winston, New York, NY T175.5M6

Morowitz, Harold J. 1993.
Entropy and the Magic Flute.
Oxford University Press, New York, NY 224 p. Q173.M87

Mumford, Lewis. 1934.
Technics and Civilization.
Harcourt, Brace and Co., New York, NY 495 p. CB478M84

Murphy, J. G., and Coleman, J. L. 1984.
The Philosophy of Law: An Introduction to Jurisprudence.
Rowman & Allenheld, Totowa, NJ 294 p.K231.M87

Ortega y Gasset, Jose. 1944.
Mission of the University.
W. W. Norton & Co., New York, NY 94 p.LB2321.072

Ortega y Gasset, Jose. 1989.
The Question of Modernity.
Prisma Institute, Minneapolis, MN 387 p. B4568.074O733

Oxford English Dictionary. 1989.
Oxford University Press.
Oxford, England 20 vol. PE1625.O87

Pagels, Heinz. 1988.
The Dreams of Reason: The Computer and the Rise of the Sciences of Complexity.
Simon & Schuster, New York, NY 352 p. QA76.9C66.P34

Pareto, Vilfredo. 1935.
The Mind and Society.
Harcourt, Brace & Co., New York, NY 4 vol. HM59.P25

Parkinson, Claire. 1985.
Breakthroughs: A Chronology of Great Achievements in Science and Mathematics.
G. K. Hall, Boston, MA 576 p. Q125.P327

Pines, David. 1988.
Emerging Synthesis in the Science.
Addison-Wesley Publishing Co., Redwood City, CA 237 p. Q101.E49

Pirsig, Robert M. 1974.
Zen and the Art of Motorcycle Maintenance.
Morrow, New York, NY 412 p. CT275.P648A33

Poincare, Henri. 1902. 1905. 1952 (tr).
Science and Hypothesis.
Dover, New York, NY 244 p. Q175.P75

Popper, Karl R. 1959.
The Logic of Scientific Discovery.
Basic Books, New York, NY 479 p. Q175.P86

Popper, Karl R. 1983.
Realism as the Aim of Science.
Rowan and Littlefield, Tatowa, NY 420 p. Q175.P864

Portnoy, Julius. 1954.
The Philosopher and Music.
The Humanities Press, New York, NY 268 p. ML3845.P7

Pynchon, Thomas. 1973. 1987.
Gravity's Rainbow.
Penguin Books, New York, NY 760 p. PS3566.Y55G7

Quine, W. V. O. 1980.
From a Logical Point of View.
Harvard University Press, Cambridge, MA 184 p. BC71.Q48

Rand, Ayn. 1971.
The Romantic Manifesto: A Philosophy of Literature.
New American Library, New York, NY 160 p. PN45.R35

Reese, William L. 1980.
Dictionary of Philosophy and Religion.
Humanities Press, Atlantic Highlands, NJ 644 p. B41.R43

Rhodes, Donald R. 1974.
Synthesis of Planar Antenna Sources.
Clarendon Press, Oxford, England 210 p. TK7872.A6R5

Rose, Frank. 1984.
Into the Heart of the Mind.
Harper & Row, New York, NY 209 p. Q335/R65

Rosenberg, Alexander. 1988.
Philosophy of Social Science.
Westview Press, Poulder, CO 218 p. H61.R668

Rosenberg, David and Brown, S. B. 1972.
The Realm of Science.
Touchstone Publishing Co., Louisville, KY 21 vol. Q111.R38

Rowell, Lewis. 1983.
Thinking About Music: An Introduction to the Philosophy of Music.
The University of Massachusetts Press, Amherst, MA 288 p.
ML3800.R6

Runes, D. D. 1968.
Philosophy for Everyone.
Philosophical Library, Inc., New York, NY 148 p. B72.R75

Sadia, Stanley. Editor. 1980.
The New Grove Dictionary of Music and Musicians.
Macmillan Publishers, London and Washington, DC 20 vol.
RL100.N48 vol.18

Sampford, Charles. 1989.
The Disorder of Law.
Basil Blackwell, Oxford, England and New York, NY 302 p.
K370.S26

Schilpp, Paul A. 1963.
The Philosophy of Rudolf Carnap.
Open Court Publishing Co., Chicago, IL 1088 p. B945.C164S3

Seligman, Ben B. 1970.
Molders of Modern Thought. A New York Times Book.
Quadrangle Books, Chicago, IL 365 p. AZ361.54

Senge, Peter M. 1990.
The Fifth Discipline.
Currency & Double Day, New York, NY 424 p. HD58.9S46

Shlain, Leonard. 1991.
Art and Physics.
William Morrow & Co., New York, NY 480 p. N70.S48

Simon, Herbert A. 1969.
The Sciences of the Artificial.
The MIT Press, Cambridge, MA 123 p. Q175.S564

Simon, Herbert A. 1991.
Models of My Life.
Basic Books, New York, NY 415 p. HB119.S47A3

Singh, M. G. 1987.
System and Control Encyclopedia.
Pergamon Press, Oxford, England 8 vol. 5192 p. QA402.S968

Solo, Robert A. 1975.
Organizing Science for Technology Transfer Economic Development.
Michigan State Univ. Press, E. Lansing, MI 224 p. HC60.S638

Spencer, Herbert. 1860. 1869. (Edited and compiled by F. Howard Collins)
An Epitome of The Synthetic Philosophy.
Williams and Norgate, London, England 571 p. B1655.C7

Spencer, Herbert. 1897.
A System of Synthetic Philosophy.
D. Appleton and Company, New York, NY 10 vol. HM51.S79
B1652A21880a

Spengler, Oswald. 1932. 1976.
Man and Technics. CB478.S6513
Alfred A. Knopf, New York, NY 104 p. CB19.S741

Springer, Sally P., and Deutsch, Georg. 1981. 1985 (rev.).
Left Brain, Right Brain.
W. H. Freeman and Co., New York, NY 320 p. QP385.5S67

Steinbeck, John. 1952.
East of Eden.
Viking Press, New York, NY 488 p. PS3537.T3234E3

Stewart, Dugald. 1792. 1971.
Elements of the Philosophy of the Human Mind.
Garland Publications, New York, NY 569 p. BF111.S7

Stott, Carole. Editor. 1991.
Images of the Universe.
Cambridge University Press, Cambridge, England 237 p.
QB43.2.I43

Suh, Nam P. 1990.
The Principles of Design.
Oxford University Press, Oxford, England 401 p. TA174.S89

Sun, H. H. 1967.
Synthesis of RC Networks.
Hayden Book Co., New York, NY 150 p. TK454.2.S9

Teilhard de Chardin (videorecording). 198x.
Synthesis, the Evolutionary Views of Teilhard de Chardin.
The Association for the Promotion of the Synthesis of Knowledge.
The Dallas Teilhard Association. 105 minutes, 3/4 in. sound.

Teilhard de Chardin, Pierre. 1959. (Bernard Wall, translator)
The Phenomenon of Man.
Harper, New York, NY 318 p. BD512.T413

Teilhard de Chardin, P. 1983.
Teilhard and the Unity of Knowledge.
Georgetown University Centennial Symposium.
Paulist Press, New York, NY 172 p. B2430.T374T37

Thompson, John. 1982.
Interaction and Coevolution.
J. W. Wiley & Sons, New York, NY 179 p. QH371.T49

Thomson, Geodrey. 1954.
A Modern Philosophy of Education.
George Allen & Unwin, London, England 283 p. LB775.T47

Todd, A. R. (ed.) 1956.
Synthesis. By R. B. Woodward. In Perspectives in
Organic Chemistry.
Interscience Publ., New York, NY 527 p.QD245.T6

Toulmin, Stephen. 1960.
The Philosophy of Science.
Harper & Row, New York, NY 176 p. Q175.T65

UNESCO. 1967.
Science and Synthesis.
Springer-Verlag, New York, NY 206 p. Q175.S418

Urmson, J. O. 1960.
The Concise Encyclopedia of Western Philosophy and Philosophies.
Hutchinson & Co., London; Hawthorn Books, New York, NY 431 p.
B41.U7

Van Doren, Charles. 1991.
A History of Knowledge.
Ballantine Books, New York, NY 422 p. AZ221.V36

Van Trees, Harry L. 1962.
Synthesis of Optimum Nonlinear Control Systems.
MIT Press, Cambridge, MA 102 p. TK7835.V3

Veblen, Thorstein. 1921. 1934.
The Engineer and the Price System.
Viking Press, New York, NY 169 p. HC106.3V39

Venturi, Robert. 1966.1977.
Complexity and Contradiction in Architecture.
Museum of Modern Art, New York, NY 132 p.
Distributed by N. Y. Graphic Society, Boston, MA NA2760.V46

Von Bertalanffy – see Bertalanffy

Vincenti, Walter G. 1990.
What Engineers Know and How They Know It.
John Hopkins Press, Baltimore, MD 326 p. TL515.V44

Wagschal, Peter H., and Kahn, Robert D. Editors. 1963. 1979.
R. Buckminster Fuller on Education.
University of Massachusetts Press, Amherst, MA 192 p. LB885.F84R2

Ward, Spencer A., and Reed, Linda. 1983.
Knowledge Structure and Use: Implications for Synthesis and Interpretation.
Temple University Press, Philadelphia, PA 681 p. BD175.K63

Whewell, William (1794-1866). 1967 (Third edition).
The Philosophy of the Inductive Sciences.
Cass, London, England 2 vol.Q175.W55 1847

White, James B. 1973.
The Legal Imagination.
Little, Brown and Co., Boston and Toronto 986 p. KF250.W45

White, Kerr L. 1988.
The Task of Medicine: The Dialogue at Wickenburg.
Henry J. Kaiser Family Foundation, Menlo Park, CA
R727.W48

White, Morton G. 1950.
The Analytic and Synthetic: An Untenable Dualism.
In S. Hook. 1950.

White, Morton G. 1955.
The Age of Analysis: Twentieth Century Philosophers.
Books for Libraries Press, Freeport, NY Reprinted in 1970 by Houghton Mifflin Co., Boston, MA 253 p. B804.A1 W5

Whitehead, Alfred N. (1861-1947). 1925. 1949.
Science and the Modern World.
Macmillan, New York, NY 304 p. Q175.W65

Wilson, E. O. 1975.
Sociobiology: The New Synthesis.
Belkap Press of Harvard University Press, Cambridge, MA
697 p. QL775.W54

World Book Encyclopedia. 1990.
World Book Co., Inc., Chicago, IL 21 vol. AE5.W55

PART 10.2
JOURNALS

American Council on Education. 1985. America's Business Schools: Priorities for Business Higher Education Forum, Washington, DC 24 p.

Arnison, G., et al (total of 173 authors). 1986. Intermediate-vector-boson propertie... Europhys. Lett. 1(7):327-45. April 1.

Billington, David P. 1986. How engineering fits our society's fabric. Agricultural Engineering 67(6):12-14. Nov/Dec.

Blakeslee, Sandra. 1991. Brain yields new clues on its organization for language. New York Times, Sept. 10 (Tues.) p. C-1, C-10.

Bordogna, Joseph. 1990. Manufacturing and engineers' education. Issues in Science and Technology 7(1):20.

Bordogna, Joseph. 1992. The Engineer's task: concurrent integration. The Bent of Tau Beta Pi 83(3):26-27. Summer.

Bower, Bruce. 1992. Reading the code, reading the whole. Science News 141(9):138. Feb 29.

Branscomb, L. M. 1986. Science in 2006. American Scientist 74(6):650-658. Nov.Dec.

Broome, Taft H., Jr. 1984. Should engineers hold the welfare of the public to be paramount?(have copy presented at AAAS Annual Meeting, May 27, 1984) 15 p.

Broome, Taft H., Jr. 1985. Engineering the philosophy of science and technology. Metaphilosophy 16(1):47-56. Jan.

Bustad, L. K., et al. 1976. Comparative medicine: progress and prospects. J. Am. Vet. Med. Assoc. 169(1):90-105.

Bustad, L. K. 1983. Dean's responsibilities and bioethics. J. Vet. Med. Educ. 9(3):97-78.

Cipra, B. A. 1993. Fermats's last theorem. Science 259:895. Feb. 12.

Clearwater, S. A., Huberman, B. A., and Hogg, T. 1991. Cooperative solution of constraint satisfaction problems. Science 254(5035):1181-83.

Conant, M. 1970. Systems analysis in appellate decision-making process. 24 Rutgers L. R. 293.

Constance, Lincoln. 1964. Systematic biology-an unending synthesis. Taxon 13(8):257-273. Nov.

Constanza, Robert, et al. 1993. Modeling complex ecological economic systems, BioScience 43(8):545-555.

Denning, Peter J. 1991. Beyond formalism. American Scientist 79(1):8-10. Jan.-Feb.

Dey, Avi. 1991. Principles of duality. TEIN, Vienna, VA Dec. 25.

Drucker, Dan. 1987. Fundamental research into the man-mde world. Mechanical Engineering 109:56-60. April.

Frosch, Robert A. 1984. On the nature of engineering research. Robert Henry Thurston Lecture. American Society of Mechanical Engineers. Dec. 13. 19 p.

Gee, Henry. 1992. Peers slam peer review. Nature 355(6360):488. Feb. 6.

Gilfillan. 1848. Tait's Magazine 15: 519.

Gillis, Anna M. 1993. As good as blood. BioScience 43(8):517-520. Sept.

Greene, R. A. 1991. Synderesis, the spark of conscious, in English Renaissance. J. History of Ideas 52:195-219. Apr./June.

Guy, Louis L. 1987. Educating engineers for the 21st century. Engineering Times 9(1):6-7. Jan.

Hall, Carl W. 1981. 1994. Chapter 16, Engineering contributions to the built environment, pp. 181-189, In "The Built Environment," edited by Tom J. Bartuska and Gerald L. Young. Crisp Publications, Los Altos, CA 380 p.

Hall, Carl W. 1988. Role of biology in U. S. engineering education. Dean's Distinguished Lecture, University of California-Davis, May 27. 14 p. [Also, ASEE Event 2204 Paper, Reno, Nevada]

Hall, Stephen S. 1990. James Watson and the search for biology's 'Holy Grail.' Smithsonian 20(11):41-49. Feb.

Hamerton, Philip G. 1863. The salon of 1863. Fine Arts Quarterly Review 2:225-262. May-Oct.

Harwood, Richard R. 1991. Education for creative synthesis against the folly of specialist power. AAAS Annual Meeting, Washington, DC. Feb. 17 p.

Hazelrigg, George A. 1991. Engineering design and decision-making. (draft of a paper) Personal correspondence. 30 p.

Herburt, G. H. 1959. The analytic and synthetic. Philosophy of Science 26(2):104-113.

Hoberman, Chuck. 1992. In Discover 13(3):70-77. Mar.

Hoffmann, Roald. 1991. In praise of synthesis. American Scientist 79(1):11-14. Jan-Feb.

Hoffmann, Roald. 1993. How should the chemists think? Scientific American 268(2):66-73. Feb.

Hughes, Robert. 1991. Lines that go for a walk. Time 138(18):96-97. Nov. 4.

Hughes, Thomas. 1989. America's real founding fathers. U. S. News & World Report 108(18):64. May 8. (Based on 'American Genesis' by Thomas Hughes)

Jerome, Richard. 1991. Honest Abe's DNA. The Sciences 31(4):7. July/Aug. (New York Academy of Sciences)

Kash, Don E. 1991. The U. S. Dilemma--Moving R & D from the laboratory to the market. NCAR 44: Southwest Research Institute, San Antonio, TX 168 p.

Kellogg, C. Gary. 1987. Analysis vs. synthesis. Engineering Times 9(1):4. Jan.

Landman, Otto E. 1993. Inheritance of acquired characteristics revisited. BioScience 43(10):696-705.

Lenhoff, Howard M. 1987. Is the organismic biologist an endangered species? BioScience 37(4):244. Apr.

Lodge, R. C. 1938. Comparison or Synthesis. Journal of Philosophy 35:432-440.

Mankin, Kyle R. 1991. Expert systems as a tool for synthesis. Graduate Seminar in Agricultural Engineering, The Ohio State University. Dec. 10. 5 p.

McKelvey, John P. 1985. Science and technology: the driven and the driver. Technology Review 88(1):38-47. Jan.

Miller, Michael. 1990. Computer tests authenicates Shakespeare. Washington Post, Apr. 21, p. C3.

National Science Board. 1988. The role of NSF in economic competitiveness. Washington, DC. 17 p.

Nealon, John G. 1985. The three arts of education. The Washington Post, p. A-22. Jan. 25.

Newcomb, L. H., and Trefz, M. K. 1987. Levels of cognition testing and student assignments in the College of Agriculture. The Ohio State University. Paper presented at the Central Region, 41st Annual Research Conference, Chicago, IL10 p.

Oliwenstein, Lori. 1993. Liquid assets. Discover 14(9):34-35. Sept.

Pedrycz, Witold. 1990. Fuzzy systems: analysis and synthesis from theory to applications. Int. J. General Systems 17(2-3):139-156. [Gordon and Breach]

Rigney, D., and Barnes, D. 1980. Patterns of interdisciplinary citations in the social sciences. Social Science uarterly 61(1):114-127. June.

Petroski, Henry. 1991. On the backs of envelopes. American Scientist 79(1):15-17.

Sinclair, George. 1977. A call for a philosophy of engineering. Technology and Culture 18(4):685-689. Oct.

Stewart, Ian. 1993. Thousand-page proof vindicates Fermat. New Scientist 139(1880):14. 3 July.

Stix, Gary. 1992. Ballpark physics. Scientific American 266(4):145-1246. Apr.

Stone, Judith. 1988. Velcro: the final frontier. Discover 9(5):82-84. May.

Tauber, A. I. 1991. Notes on the Neanderthal. BioScience 41(6):416-418. June.

Time Magazine. 1956. Math prodigies. Time 67(12):83. Mar. 19.

Time Magazine. 1973. Ten best fiction books. Time 102(24):56. Dec. 31

Vaughn, C. 1988. Shroud of Turin is fake. Science News 134(14):229. Oct. 1.

Wallis, Claudia. 1991. Why new age medicine is catching on. Time 138(1):68-76. Nov. 4.

Walters, Eric A. 1988. Professional, creative leadership: The origins and development of the Carnegie Plan. A thesis for partial fulfillment of honors graduation, College of Humanities and Social Science, Carnegie-Mellon University, Pittsburgh, PA. 31 p.

Wang, Daniel. 1992. Bioprocess engineering in Japan. JTEC Panel, National Science Foundation, Wshington, DC May, p. 86, 198 p.

Waters, Tom. 1992. Unfolding world of Chuck Hoberman. Discovery 13(3):70-77. Mar.

White, Robert M. 1986. Future imperfect. Presented at the Electric Power Research Institute, Monterey, CA. Aug. 19. 28 p. [Also see EPRI Journal 12(2):20-25]

White, Robert M. 1987. Technological change — a global current in flood. Lecture at Case Western Reserve University, Cleveland, OH, Sept. 22. 15 p.

White, Robert M. 1989. The engineer and the changing business world. Siemens Review 56(1):38-41. Jan./Feb.

Wilson, Kenneth G. 1991. Personal discussion on synthesis and design of an education system. The Ohio State University, Columbus, OH. Dec. 3.

Winkler, K. 1987. Interdisciplinary research. Chronicle of Higher Education. Oct. 7. p. A-1, 14-15.

PART 10.3
SERIALS RELATED TO SYNTHESIS

Analysis-A journal of philosophical writings. 1933-
Basil Blackwell, Oxford and New York. (Formerly Barnes & Noble)
Four issues per year. B1.A11

Analysis-vol. 1/2, 1933/35; 1967-
Barnes & Noble, New York, NY B1.A112

Analysis and Synthesis. 1956-
International Center for Free Trade Unionists in Exile. DR1.A5

Analyst/The Analyst – numerous periodicals by this name

Configuration. 1993-
Published by the Johns Hopkins University Press for the Society for Literature and Science in cooperation with the Georgia Institute of Technology.
Three issues per year. Z39.48

Integrated, Integration, Integrative – numerous periodicals by this name

Interdisciplina.
SLM Press, Tallahassee, FL
Four issues per year. AP2. I 643

Interdisciplinary Science Reviews. 1976-
Heyden, London
Four issues per year. Q1. I 577

Interface – numerous periodicals by this name

Interface, the computer education quarterly. 1979-
Santa Cruz, CA
Four issues per year. QA76.27 I 56

Interface (Carmel). 1978-
International Computer Programs, Inc.
Carmel, IN .QA76.9.M3 I 57

Interfaces. 1971-
Institute of Management Sciences.
Operations Research Society of America HD28. I 45

International Journal of General Systems. 1974-
Gordon and Breach Science Publishers, New York, NY
Four issues per year. Q295.I58

Interstudy Bulletin — The newsletter of the International Association for the Study of Interdisciplinary Research.
Graduate School of the University of Washington, Seattle, WA.
Three issues per year.

Journal of Philosophy. 1904-
New York, NY. B1J6

Metaphilosophy. 1970-
Basil Blackwell, Oxford, England.
Four issues per year. B1M46

Methodology and Science. 1968-
International journal for the empirical study of the foundations of science and their methodology.
Esser Scientific Press, Haarlem, The Netherlands.
Four issues per year.

Multidisciplinary Research (psychology)
International Multidisciplinary Association
Two issues per year. BF11. I 615

Parabola. 1976-
Parabola, 656 Broadway, New York, NY
Four issues per year. BL1.P25

Philosophy and Phenomenological Research. 1940-
by the International Phenomenological Society.
Brown University, Providence, RI
Four issues per year. B1.P57

Philosophy of Sciences. 1934-
Journal of the Philosophy of Science Association.
Department of Philosophy, Michigan State University, East Lansing, MI
Four issues per year. Q1.P55

Proceedings of the Aristotelian Society for the Systematic Study of Philosophy (PAS). 1900-
Basil Blackwell
Annual. B11.A47

Synergy, San Francisco Bay area reference center, 1967-
San Francisco, CA
Monthly. Z671.S95

Synthese — An international journal for epistemology, methodology and philosophy of science, devoted to present-day cultural and scientific life. 1936-
Kluwer Academic Publishers, Dordrecht, The Netherlands.
Monthly. AP1.S9

Synthesis — An undergraduate journal in the history and philosophy of science
Mathar Press, Cambridge, MA
Four issues per year. Q124.6.S92

Synthesis — An international journal of methods in synthetic organic chemistry. 1969-
Georg Thiema Verlag, Stuttgart and Academic Press, New York, NY.
Twelve issues per year. QD262.S93

Synthesis and Reactivity in Inorganic and Metal-Organic Chemistry. 1971-
Marcel Dekker, Inc., New York, NY
Ten issues per year. QD156.S9

Synthetic Communications — An International Journal for Rapid Communication of Synthetic Organic Chemistry. 1971-
Marcel Dekker, Inc., New York, NY
Twenty two numbers in 1990. QD262.S935

Systems – numerous serials are published with the word systems in the titles, several of which contribute to a better understanding of synthesis

Part 11.
Copyrights and Permission to Use Materials

Extension of the Copyright Page.

The table, titled 102 Ideas, reprinted from **The Great Ideas.** Mortimer J. Adler ©1952, 1986. By permission of Encyclopaedia Britannica, Inc., Chicago, IL.

The 4 Dimensions of Philosophy by Mortimer J. Adler, ©1993. A quote from page 258. Macmillan Publishing Company, New York, NY.

©Quote of Roberto Ardigo on page 26 of **Dictionary of Philosophy and Religion** by William L. Reese. 1980. Reprinted with the permission of the publisher, Humanities Press International, Inc., Atlantic Heights, NJ.

pages 12, 73, and 290 in the **Dictionary of Philosophy** by Peter A. Angeles, ©1981. Reprinted by permision of HarperCollins Publishers, New York, NY.

page 4 in **Creativity: The Magic Synthesis** by Silvano Arieti, ©1976. Reprinted by permission of HarperCollins Publishers, New York, NY.

©Isaac Asimov, **Asimov on Science.** Permission is granted by Bantam, Doubleday, Dell Publishing Group, Inc., New York, NY to quote from page 391. 1989.

Table 1-1, page 117 (adapted) from **The Coming of Post-Industrial Society** by Daniel Bell. Copyright ©1973 by Daniel Bell. Reprinted by permission of HarperCollins Publishers, Inc., New York, NY.

David Berlinski ©1976. Permission to reprint Table 3.1, page 127, from the book, **On Systems Analysis,** by The MIT Press, Cambridge, MA.

Copyright ©1987, by Allan Bloom, **The Closing of the American Mind**, for an excerpt on page 228 of the book. Reprinted by permission of Simon & Schuster, Inc., New York, NY.

©**Miracle at Philadelphia** by Catherine Drinker Bowen. 1966. Little, Brown and Company, Boston, MA.

Chuck Hoberman, ©1992. The Walt Disney Co., Discover Syndication. Reprinted with permission of **Discover Magazine** to quote from the March 1992 issue.

©Raold Hoffmann, permission to quote from In Praise of Synthesis, **American Scientist** 79(1):11-14 (1991).

Scientific American, ©1993. For excerpts from an article, How should chemists think, by Raold Hoffmann, in Scientific American 268(2):66-73.

©Gerald Holton, **Thematic Origins of Scientific Thought.** 1988. Published by Harvard University Press, Cambridge, MA for matrial quoted from page 96.

pages 38 and 39 in **Perpetual Innovation** by Don E. Kash, ©1989. Reprinted by permission of HarperCollins Publishers, New York, NY.

©C. Gary Kellogg, Analysis vs. Synthesis. Engineering Times v. 9, n. 1, page 4, 1987. Reprinted with the permission of **Engineering Times,** National Society of Professional Engineers, Alexandria, VA.

©Rupert C. Lodge, **Philosophy of Business.** The University of Chicago Press, Chicago, IL for use of the illustration, 'philosophy of business.' 1945.

Robert Lodge, **The Great Thinkers.** ©1949 and 1968. Excepts from page 196 and 252 of the book. Reprinted by permission of Frederick Ungar Publications, of the Crossroad Publishing Company, New York, NY.

James A. Michener, **Journey,** for use of a quote on page 19 of the book, ©1989, Random House, Inc., New York, NY.

An excerpt from the book, **A System of Logic** by J. S. Mill, 1843. Reprinted by permission of HarperCollins Publishers, New York, NY.

©Arthur I Miller, **Imagery in Scientific Thought.** Birkhauser, Cambridge, MA for use of Fig. 1.1 Poincare's epistemology, page 29. 1984.

Permission granted by **John G. Nealon,** Fairfax, VA for use of his letter to the editor of the Washington Post, January 215, page A-22. 1985.

Permission granted by **L. H. Newcomb,** The Ohio State University, Columbus, OH, for use of his table on cognitive levels and a figure on taxonomy related to learning. 1987.

Jose Ortega y Gasset, **Mission of the University,** trans. with an introduction by Howard Lee Nostrand. Copyright 1944, ©renewed 1972 by Princeton University Press. Used quotes from pages 40, 78, and 80 with the permission of PUP.

©Heinz Pagels. 1988. **The Dreams of Reason.** Simon & Schuster, New York, NY. Permission to use quote on page 36 granted by Elaine Pagels.

©Withold Pedrycz, 1990. Permission granted by **Gordon and Breach Science Publishers** for use of Figure 1, visualization of variability of the cognitive perspective in case of one input-one output of a fuzzy system, in the International Journal of General Systems, v. 17, n. 2-3, pages 139-156.

©Henry Poincare, **Science and Hypothesis.** 1952. Permission is granted by Dover Publications, New York, NY to quote J. Lamor from the introduction on page xix.

Charles Sampford ©1989. **The Disorder of Law.** Permission to use Figure 5.2 granted by Blackwell Publishers, Oxford OX4 1JK, UK.

©Peter Senge, **The Fifth Discipline.** Doubleday, a Division of Bantam, Doubleday, Dell Publishing Group, Inc., New York City, NY, to use the unnumbered illustration, 'systems thinking,' page 375. 1990.

©**The Science of the Artificial** by Herbert A. Simon. 1969. A quote from page 118. The MIT Press, Cambridge, MA.

An excerpt from pages 257 and 258 of the book, **Models of My Life** by Herbert A. Simon, ©1990. Reprinted by permission of HarperCollins Publishers, New York, NY.

From **East of Eden** by John Steinbeck. ©Copyright 1952 by John Steinbeck, ©renewed in 1980 by Elaine Steinbeck, John Steinbeck IV and Thom Steinbeck. Used by permission of Viking Penguin, a division of Penguin Books USA Inc. for North American rights, and Reed Book Services, Ltd, a member of the Reed Elsevier group, Rushden, Northants, England, for the remainder of world rights.

Carole Stott ©1991, **Images of the Universe,** for use of the figure on 'unification scheme of the universe,' on page 233, reprinted with the permission of Cambridge University Press.

©1991. American Institute of Biological Sciences. Alfred E. Tauber, author article in **BioScience** 41(6):427-428.

©Pierre Teilhard de Chardin, **The Phenomenon of Man.** ©1955 by Editios de Seuil and ©1959 in the English translation by William Collins Sons & Co., Ltd., London and Harper & Row, Publishers, Inc., New York, NY for permission to quote.

Quote on page v from **Interaction and Coevolution** by John Thompson. ©1982. Permission granted by John Wiley & Sons, Inc., New York, NY for material on page 21 and pages 267-268.

Quote from page iv of the book, **The Age of Analysis.** Copyright ©1955 by Morton G. White. Reprinted by permission of Curtis Brown, Ltd., New York, NY.

©**Science and the Modern World** by Alfred North Whitehead. 1925. 1949. An excerpt from pages 151 and 152. Macmillan Publishing Company, New York, NY.

©McGraw-Hill, Inc., New York, NY for use of material from **An Introduction to Scientific Research** by E. Bright Wilson. 1952.

Various styles have been used to give credit for the permissions provided by a number of individuals, publishers, and organizations. These styles represent the manner in which credit for permissions were suggested or requested.

Part 12.
Elements (Index)

A

B

D

E

F

Q

R

S

T

U, V, W

X, Y, Z

Part 13.
Ending

"Great is the art of beginning,
But greater is the art of ending."

Henry Wadsworth Longfellow

"This may be for the book an ending,
But generate for synthesis a new beginning."

Carl W. Hall

THE AUTHOR

The author lives with his wife, Mildred, in Arlington, Virginia 22207-1033 where he is a consultant. His numerous books and articles cover drying, energy, engineering, and related topics. He is a member of the National Academy of Engineering (N.A.E.) and is a registered Professional Engineer (P.E.) and has been active in numerous engineering organizations. He retired from the National Science Foundation (N.S.F.) in 1990; previously he was Dean of the College of Engineering, including Architecture and Environmental Sciences, at Washington State University, 1970–1982. He obtained degrees and was on the faculties of the Ohio State University, University of Delaware and Michigan State University, following service in the infantry in WWII. He completed the SMG Program at JFK School at Harvard.

Printed by
CPI books GmbH, Leck

FSC
www.fsc.org
MIX
Papier | Fördert gute Waldnutzung
FSC® C083411

Kolibri 360 GmbH
Ferdinand-Jühlke-Straße 7
99095 Erfurt, Deutschland
produktsicherheit@kolibri360.de